NATURVORE POWER

Transcending Industrial Food and Medicine

BY

CHARLES C. HARPE, MD

Published by Warren Publishing, Inc.
Huntersville, NC
www.warrenpublishing.net

ISBN: 978-0-9884170-2-1

Library of Congress Control Number: 2013933928

I am indebted to a most excellent family, especially Claudia, whose endless patience and support allowed me to persevere. I am so thankful to Dad, who provided me a healthy example and made me understand that my choice of words really does matter. To my friend Larry, thanks for the generous encouragement despite being tortured by the convoluted sentences from this novice writer. Thank you Wendy for the encouraging photo sessions and your keen photographic eye (www.wendynewmanphoto.com).

THIS BOOK IS FOR MOM (1934-2012).

PREFACE

Life evolves from our choices and the cover photo illustrates this point. I am having more fun with my health than at any other time in my life, even to the point that I consider health to be the ultimate form of self-expression. To be able to run up the Blue Ridge Mountains with my teenagers, snowboard, bodybuild in any capacity and free dive the coral reefs in Hawaii and the Florida Keys enhances the meaning of my life. Although not often considered meaningful in an existential sense, our choices of food and activities tend to either improve us (tissue repair or hypertrophy) or destroy us (tissue corruption or atrophy). There is no middle ground in the physical world (reality), yet we are prone to believe that we can coast through life, relaxing and enjoying a static biological state as if it were real. Within these pages, I intend to dispel the notion of health being either a given in life or associated with a transaction of medical care. Both fallacies are actually part of a complex delusion born of human calculation and rationalization. To be real, as in really healthy, we must consider our choices for they demonstrate our sincerity in matters of life.

I am the man pictured on the cover of this book, but I never previously enjoyed the physical condition I have now. The first forty years of my life were spent creating a typically successful American life, accumulating achievements, fashioning a worthy career (which included board certification in internal medicine), raising a family, building an executive home, and contributing to a 401(k). But within that American dream, my identity also acquired some negative attributes, such as 55 pounds of excess body fat, along with a weaker form and functional capacity. Insidiously and unexpectedly, I had become complacent while adopting the new American "normal" of physical attributes and melding into the unhealthy herd of "successful" people. However, one day I finally took notice of my seasoned identity and this provoked a powerful change within me: I realized I could express myself differently by choosing my foods and activities more consciously. Indeed, by virtue of some mundane actions, I have since improved my essence -- that being the tissue between my skin and my core, the foundation of my entire existence. Now, I am ready for anything, all the time. Armed with this new insight, it is difficult to view the world as I once did.

This book not only records the initial stages of my journey toward self-improvement, but is also an offering, a metaphorical fruit, perhaps one that ripens best with a bit of self-reflection. Upon such reflection, it is now clear that I had spent the first four decades of my life detached from an essential part of my humanity -- my self-preservation. I had become someone that I did not wish to be. How could I have been so oblivious to the gradual loss of essential personal qualities, such as those that make me well? The answer to this interesting question became clear once I cultivated the expression of a more natural and more functional me. Within these pages I explain how I became a Naturvore.

Introduction

I designed this book to help other people undergo a transformation to a healthier state and a more meaningful life. To accomplish such a transformation and remain healthy, one must have insight into their internal (biology) and external (society) circumstances. With such an insight, we will clearly see that American culture fashions us, by creating both our internal and external circumstances, into people we do not wish to be -- obese and sick. We can choose to be healthier, happier and more productive, but our success will be determined through our relationships to two orders of operations, the Industrial Order and the Natural Order. Our relationship to these entities is reflected in our choices.

I divided the book into two sections, the Industrial Order and the Natural Order, so that my readers can identify two diametrically opposed forces that influence their fates. As young Americans, we typically develop an industrial identity that permeates us all the way to the cellular level (largely by manufactured foods), but the medical-industrial complex not only enhances this unnatural identity, it also capitalizes on the cognitive dissonance and mental twists created out of our physical corruption. Within the section named *Industrial Order*, I explain our propensity for sickness and within the section named *Natural Order,* I explain health through natural means.

This book was made possible only after my body was restored to a more authentically natural state, the Naturvore identity. Meaning that by simply switching my foods to those provided whole from nature, I not only reversed the corruption of my natural identity and reinstated my primordial wisdom (the body's wisdom), I felt improved consciousness which culminated into this book. As a physician/ scientist, this was an amazing experience in that my natural identity which had been corrupted by the Industrial Order was reinstated in the reverse manner through which it had been lost, simply by changing my diet. This realization was unexpected and extremely atypical for a highly trained soldier of the Industrial Order. Had I not experienced the reversal of my industrial identity to a more natural identity for myself, I would not believe my own story (perhaps even more improbable is that I am broadcasting such a concept). Very importantly for those who want health, the sequence of events spelled out in the book is no accident, nor a human innovation, in that it simply follows my personal life. Everything happened as it was intended, naturally.

My personal habits are disclosed for sincerity, credibility and to instill optimism. I want every reader to understand that what they eat, what they do and what they believe is important in fulfilling the ultimate role of their life -- that of a healthy organism. By the time my readers reach *Transcendence*, they will surely understand that "life is what you make it," an optimistic posture that can be realized on a genetic level and even an atomic level. We can lose a great deal of excellent life, the unrealized and unmeasured life, while corrupted by the food, acts and beliefs created by the Industrial Order.

I have kept scientific explanations brief so as to not burden readers with too much biochemistry or medical jargon, perhaps enticing them to search for more. These tidbits of biological science should stimulate a bit of awe for the biological events we take for granted, specifically those that lie below our level of consciousness. This information also serves to defend my positions on important matters of health; the reader's appetite for reductionist explanations is expected.

I discuss my professional dilemmas, not to be mean-spirited or overly critical of the medical profession, but to instead provide insight into how the medical-industrial complex affects all of us. It not only devours important national resources, but also corrupts our minds, and our beliefs about health, whether we are patients or not. I created vignettes of patient encounters that are composites of real encounters over the years; their discussion adds life and a more personal perspective to the various points to ponder. The patient encounters I describe are indeed typical and I suspect some people will identify with these scenarios without ever being my patient.

I have chosen not to complicate my writing with listing references within the text. My goal was to strike a more conversational approach, as if I were speaking to a patient in my clinic or as if the reader were my friend. The information I express is backed by science, but is not as complete or referenced as one might expect within a medical reference or a textbook on nutrition. This is simply my story, while providing some of the information I use to assure my own health.

There are no "standards of care" presented in this book, as are found in modern medical care, therefore no one should feel compelled to live or perform exactly as I do. Dissent and deviation from my plan are both expected and tolerable because the reader will ultimately decide on the level of results they wish to attain and only then will the reader's health plan be practical and personal. Our health is the most important form of self-expression we can undertake and individual results depend upon the energy (literally) that we put into it. Nevertheless, people who achieve a high level of confidence regarding their health and fitness will do so on their terms and in their time. We must find what is right for us as individuals and this is why real health is based upon self-discovery more than knowledge from others. Health is a practiced affair, not a learned affair.

There is urgency in the plight of the American people as hundreds of millions are suffering chronically. For this reason, I am blunt at times and try to maintain a sober, yet inspirational, tone throughout the book. I want people to undertake their own personal search for meaning and value within their body, that being the value of their biological integrity. Furthermore, as people tend to view life more optimistically, a more positive context unfolds in their health. We are generally bred to be healthy and this is a power given to us through the Natural Order. I choose to accept this physical reality and others can too.

THE INDUSTRIAL ORDER

ACQUISITION OF AN INDUSTRIAL IDENTITY

"In nature it is not the strongest or smartest that succeeds but the one most able to adapt."

— Darwin

My arrival as a distinct biological entity occurred on March 7, 1964. I was born in Madrid, Spain, on that day to Helen and Clifford Harpe, and it could be said that I also began my quest for an industrial identity at that same moment. My biological potential would be molded by the industrial world from that point forward, the shape of my life bearing the mark of the various industries that are part of and, to a large degree, form American culture.

Spain was not as industrially advanced as the United States in 1964, but my parents and I lived as Americans in Spain. I ate mostly processed American baby food because my father, having served in the military, was able to use the US commissary for these items; but I became a "Gerber baby" not just because I was too young for a traditional Mediterranean diet but because my mother did not trust the food prepared by the maid as much as American products when it came to nutrition for her first child.

My father was working with the United Nations High Commission for Refugees, primarily assisting refugees from Castro's Cuba seeking immigrant visas to the United States. I heard my early days discussed many times, not only by my parents but also by close Cuban friends who were living in Spain at the time. By all accounts, those days were a happy period. My parents thought life in Madrid in the mid-1960s was exciting, and in the pictures from that time everyone looks like movie stars. In my second year of life, while still a Gerber baby, we moved to Miami, Florida, where I spent the majority of my childhood and became yet more American in my diet.

My parents divorced when I was four years old, and because mom never remarried, my brother and I were raised in a single-parent home. Mom had a full-time job with the US government, and so my brother and I were shuttled from schools to, thankfully, *great* babysitters. My mother had little time to cook, given the rigors of being a working single parent, and consequently much of my nutrition came from fast foods and manufactured packaged foods. Convenience was important to my mother and, when I was old enough to fend for myself in response to hunger pangs, to me as well; but I also remember that I actually wanted fast food because that was all I knew to eat. Fresh food seemed inferior as a food source, largely because it seemed to be the old way of life, much as my mother had viewed the traditional Spanish diet when I was a baby.

The TV advertisements for my food and the packaging it came in instructed me that modern technology-created foods were better. Wheaties was the breakfast of champions, as everyone knew, and the assertion served as self-affirming proof that technology was making food better. Food that was simply grown in the ground, foodstuff that did not include the heavy hand of man in its construction, was viewed as archaic. Besides, natural whole foods are nowhere near as exciting or as tasty as processed foods in their brightly colored boxes and wrappers. That is what we were propagandized into believing.

Through the 1950s and 1960s, TV provided a new medium to reach American households, and through that magic box we were encouraged to buy more manufactured foods. It is no accident that a child is bombarded during Saturday morning cartoons with enticing advertisements. Kids became a target market not just for toys in my youth. I nagged and pleaded with my parents to purchase "magically delicious" Lucky Charms cereal, which not only looked bright and fun with its multi-shaped and multi-colored marshmallows but also came with a toy surprise in every box. Other precision-made sweet cereals could change color in milk or "snap, crackle, and pop," and Apple Jacks was much sweeter than real apples. Marketing aimed at children also gave us Trixx cereal, which "was for kids," an assertion that is appealing in and of itself somehow, and mascots like the iconic Tony the Tiger and Cap'n Crunch. These pitches became not only fixtures on Saturday morning TV but part of American popular culture. Nobody was concerned about the harm or the lack of nutritional value of heavily processed food at the time, and I don't remember having much access to fruits and vegetables, nor did I ever really crave them. Natural foods did not seem to make my taste buds dance or my brain feel intense comfort as processed foods did.

It wasn't just breakfast cereals specifically pitched to children that formed my American diet either. Other products and their catchy phrases became embedded in my young mind too. Campbell's Soup was "Mm-Mm Good" and Kentucky Fried Chicken was "Finger Lickin' Good." Oscar Meyer Bologna and Wonder Bread, according to the ads, were supposed to make us healthier and became staples of the American diet. Variety was also pitched to us as important, and it came to us constantly, on full display with bright, glossy pictures while we were promised a bargain as well in the form of supermarket coupons in the Sunday newspapers. As a child, I enjoyed reading the newspaper anyway, but the coupons promoting new food products, ads that enticed us to try the latest industrial food creations, made the Sunday paper additionally enticing.

Interestingly, I felt that my family was fortunate to be eating more foods from American factories than did my friends of other nationalities living in Miami. My Cuban friend Omar had to dine on chicken and rice, meat and rice, or bananas and rice every night. I had the luxury of newly fabricated TV dinners, Fritos, and Burger King hamburgers and fries instead. Omar drank boring old water or milk with dinner while I had Coca-Cola.

I had substantial opportunities to compare my modern American food to the traditional cuisine of another culture because the Cuban influence in Miami was in full swing at the time. From 1966 until 1978, because my father worked in the administration of the Cuban Refugee Program for the Federal Government, we welcomed the company of many Cuban acquaintances and friends. To this day my favorite Cuban food creation is the sandwich *Cubano*, a pork, ham, and cheese sandwich usually served with *platanos maduros* (fried, sweetened, ripe plantains) and *frijoles negros* (highly seasoned black beans). But for my average everyday meals when I was young, I wanted fast food fare.

In contrast to the more sterile scenes of the average fast food restaurants, where strangers eat quietly seated on plastic benches, Cuban restaurants were extra lively and, in many ways, meals there were more like a social get-together, with groups talking between tables, their voices several decibels higher than in American eateries. Everyone espoused a feeling of well-being and belonging in this adopted land, much as they would have back in Cuba. My family's favorite Cuban restaurants were the Versailles and the La Lechonera, and eating in these establishments was definitely a different cultural experience compared to eating at Burger King. Despite these excellent cultural cuisine experiences, which admittedly were not always healthy dining experiences, I believed that American ingenuity in the production of newly-created foods made modern fare somehow superior to Old World foods, while Cuban food experiences of my early years were also out of step with fast-paced American needs.

As modern Americans, we believed, given that we were all on the move, even eating should reflect our cultural predilections for consistency and efficiency. Consequently, foods made in an assembly-line fashion were readily embraced by our generation because they reflected our ideals, including the satisfaction of brand loyalty to our favorite corporations. The omnipresence of the brightly-colored institutions like Burger King, Wendy's, and McDonald's was a welcome solution to our need for speed for the busy person.

The expansion of convenience stores, strip malls, and fast food restaurant chains assured me that life was getting better everywhere. During this time in my life, there were numerous automobile trips with my younger brother to visit our grandparents. At the beginning, I expected that it would be difficult, if not impossible, to find acceptable food quality and selection along those long 14-hour car rides between Miami and Oak Ridge, Tennessee. But I was relieved when I saw that the people of Valdosta, Georgia, were able to enjoy the accoutrements of modern living because they had both a Burger King *and* a McDonald's restaurant.

As I moved through life, I was comforted by the sameness of the industrially produced food that was the heart of American cuisine. My favorite fast food choice, the Burger King cheeseburger, was the exact same no matter where I was, and so my personal identity was maintained. I appreciated the small element of individuality granted when Burger King instituted the "Have It Your Way" slogan. Wow! One could order a Whopper with or without pickles or onion or mayonnaise, or even double the meat, if desired. But this minor personal touch still meant that variety was quite limited.

At that time in my life, milk and water were both boring to me, and I had little use for either except when adding milk to breakfast cereal or getting a glass of water to quench my thirst after some strenuous exercise. I am sure I consumed more Coca-Cola products than did other kids I knew, and I even remember how happy I was to see my father bring home the first one-liter bottles of Coke, which were made of glass and so quite dangerous, especially when you dropped one on a terrazzo floor like those we had in Miami. The big bottles meant we did not have to open so many to get our fill. The commercial life span of these containers was short, giving rise to something safer and easier to handle, the even bigger and infinitely more convenient plastic two-liter bottle.

If American industry took great pains to become an integral part of our childhood lives and our childhood memories during my youth, the movement toward targeting kids as a market went into overdrive in subsequent years. The evolution of certain manufactured products, even their packaging, is stored quite vividly in my memory banks, as vividly as other childhood memories, but that was before the onslaught. Using applied psychology, fast food chains developed special meals just for children. "Happy Meals" included everything a child could want: a small cheeseburger or chicken nuggets, French fries, a small soda, and this week's toy, all in a compact package. It was sheer marketing genius that yielded entertainment and taste at a competitive price in a single product for kids. Nutrition was not the primary concern with regard to our purchases when I was growing up, and it was not for a very long time after. We trusted that these food companies had our best interests at heart. How could we not trust the athletes on the cereal boxes, the Burger King or Ronald McDonald characters pitching their respective products?

The American industrial machine, with its unrivaled ingenuity for convenience and mass production, had given us a cultural identity; and therefore it seemed so American to enjoy manufactured foods. Civility American-style was linked to our industrial production schemes. America was young, and so we had not achieved the cultural sophistication of Europe, but their methods of eating seemed stuffy and moribund to us anyway. I liked to think that modern humans were destined to be mobile, and so there was nothing appealing in the tedious preparation and consumption of foods European-style. After all, sitting down to eat several courses of a meal was not only boring but would disallow the efficiency of eating while zooming down the highway in our cars.

I distinctly remember the general mood of the 1970s as one of hope and wonder (although many readers may have bad memories of the Viet Nam War and Watergate, I was too young to be greatly affected by politics). Our fast-paced and increasingly technological culture, brought to us by industrial gadgetry, provided ample promise of a happier life simply because we were modernizing. America's bicentennial was celebrated with great pride in our democracy and in capitalism, airplane travel was becoming routine, and the imaginative among us were picturing space colonies and moon outposts.

The Apollo space program had stoked my imagination with notions of a better life ahead and caused me to contemplate space travel myself. At Disney World, Space Mountain opened the same year my father welcomed very large, sophisticated computers into his office. We had, seemingly overnight, become a busy society: building, processing, consuming, and automating our lives. Very few people considered our culture's effects upon the natural world and the planetary resources. Fewer still appreciated the effects of American industry in corrupting the natural attributes and biology of our people through our overall industrialization process and perhaps especially through our industrial food production practices.

Just as I did when I was a child, my own children expect a diet consisting of manufactured foods. Modern American children are continually shown through potent advertisements that celebrations can only occur with industrial creations such as Happy Meals; therefore they do not consider a diet of natural foods as being normal, much less desirable. A good performance (or simply showing up for the game) on the soccer field requires a fun-meal treat, for example. Even worse, such foods are offered to our children for mood control and stress relief, manufactured food used as a pick-me-up after a bad school day or a disappointing baseball game.

The rewards we give ourselves by way of manufactured foods are actually felt as rewards on a more profound level than just symbolism. Healthier, non-processed foods do not appeal to children as fantastic treats like processed sugars and fat, which stimulate the reward centers in our brains. Food scientists understand very well that there are particular areas in our brains that govern our emotions and can reinforce our behaviors: the nucleus accumbens, the ventral posterolateral nucleus, and the amygdala. Stimulation of these nerve centers located deep inside the more primitive areas of the brain is interpreted as pleasurable. The nerve centers that tend to induce a desire to repeat certain activities contain the cannabinoid receptors that are also responsible for the intoxication associated with marijuana use. Highly processed, manufactured foods have been shown to increase the number of these cannabinoid (CB1) receptors and increase receptor signaling strength.

It should be quite obvious that, as we consistently ingest processed foods over time, there will be a compulsion to consume more of the same products. Perhaps more important with respect to compulsory behaviors, the uniformity of the experience, the consistency of the particular product and therefore the consistency of its effect on the brain, is as important as the initial effect. In short, industrial consistency improves the influence of the food product's influence on a person's brain, and with consistency in design and consistency in delivery, manufactured foods tend to induce very potent effects on the brain. In fact, as the receptors' signaling strength is increased (the power of the chemical's influence on the brain) and the number of actual receptors increase with repetitive stimulation over time, there is a sensation of discomfort when consumption of these processed foods is stopped. A food craving is likely "hungry" CB1 receptors absent the stimulation by scientifically-created foods. Albeit less intense, there are distinct similarities between the way drugs and manufactured foods affect our brains. Like psychoactive medications, eating so-called highly palatable manufactured foods gives us sensations of comfort when ingested and cravings when not ingested. In short, despite the ill effects, including hyper-excitability in children, manufactured foods have been shown to induce repetitive consumption. Obviously developing food product loyalty really means developing a loyal brain chemistry reaction, a loyal consumer who is accustomed to the effects of the manufactured foods on their brain.

Food textures also affect reward centers in the brain. A noted Harvard anthropologist, R. Wrangham, found that rats fed soft processed foods grew an average of 30% more in just 26 weeks of life compared to rats fed hard re-compacted pellets. His experiments indicated that soft-textured foods (breads, cakes, pastries, and croissants) stimulate the amygdala, an area of the brain important to reward and memory, more than hard foods do. Indeed, we are comforted by our comfort foods, and that strong behavioral reinforcement provides the impetus for food that is technologically produced to be continually consumed. Our food products become part of us, part of our identity, an industrial-made identity, which is how brand loyalty works on us, no matter the product but especially with regard to food. We are, in fact, what we eat, and advertisers know it.

Food scientists toil incessantly to invent creations sold as food that control our taste-bud response and brain chemistry to enhance their market share and profits. Children are the most vulnerable population, enticed by seemingly benevolent cartoon characters and colorful boxes containing not just food but frequently also toys. We must not forget that the convenience and omnipresence of these food creations will also influence people's behaviors. As a child, I could not have known that my brain chemistry was changing every time I chose manufactured foods over natural foods. It never occurred to me that the products appearing virtually everywhere, such as in vending machines, check-out lines, theaters, convenience stores, airports, and even schools, could be harmful to my brain and body. I believed injurious products were labeled as such and stored away, that I was protected from anything ultimately harmful to my health.

Our holidays are great for industry, and now what we label a holiday appears to be determined by Hallmark. Apparently, without Grandparents' Day, Secretaries' Day, Bosses' Day, Doctors' Day and the like, we will all feel unappreciated. Many people now abhor the materialism injected into our traditional holidays, and some are beginning to realize the insanity of excess industrial food consumption associated with many holidays and celebrations. However, that has not stopped the invention of ever new ways to celebrate unhealthily. I venture to say that one holiday in particular was created out of the desire to showcase industrial splendor: the Super Bowl.

The Super Bowl, once merely a football championship game, has become a special holiday tailor-made for industrial food and booze manufacturers. On this "sacred" winter night, we as a nation get poised to critique the TV commercials for beer, chips, soda pop, fast food, and other iconic products of American industry. Personally, I can remember the commercial advertisements better than the actual football games. There is little doubt that American industry manipulates our behaviors through the sheer inundation of the coded and not-so-coded-messages of advertising, and, as noted previously, those products are woven into our biological fabric by the food and beverage industries.

We are also affected psychologically by the slick advertisements and inaccurate messages saying processed foods are healthy and natural (which means a product is assembled out of ingredients found in nature, not that it is natural in the truest sense of the word, as the reader will see in subsequent chapters). Today, the buzz within the world of processed foods is the fortification of artificial foods with vitamins and minerals, these unhealthy foods now promoted as beneficial. Likewise, the method of preparation is touted as healthy, and so, for example, instead of fried chips, we can have baked chips, although the deleterious health effects remain. In short, we are convinced that the new product is really good for us when in fact it is little changed from the old.

The industrial food companies are trying to convince us that they have our best interests in mind, simply capitalizing on the new concerns about our health of course, even though it is largely their products that are causing our nation's health problems. Corporations who dole out these inventive food products want our trust because we become loyal consumers as long as this illusion is not broken, but that loyalty is actually killing many of us. This is the overarching message of this book: you must change how you eat if you want to live a long, healthy life.

Americans consume food items that are as consistently produced as our cars, radios, telephones, medicines, and other industrial products. We are comforted by that consistency and believe our physical and emotional needs are met by these creative inventions. This is an industrial identity, an idea of self, of who we are, premised upon what we consume. The initiation into our industrial identity is through the American food chain and begins for all intents and purposes at our birth. Industrial food products are not merely dependable, however; they are part of us by affecting us psychologically and biologically, sadly, like an illicit drug habit affects the addicted. Our manufactured foods also provide us with a sense of shared culture, provide new twists to old holidays and a reason to conjure new holidays or reasons for celebration. In few words, these foodstuffs have entered us, body and soul.

Toys, movies, food, and other products of the modern age were sold to me on television and in radio advertisements, ample opportunity to influence my identity. As a late baby-boomer, I witnessed the cultural conversion from the austerity principles of the pre-World War II generation to the values of the empowered consumer, the new American identity. Brand loyalty became an important part of each person's identity, including to the foods we chose to consume. We were not just a General Motors (to this day I still will not own a Ford truck, only a Chevy) and a Schwinn bike family but a Burger King and Coca-Cola family. I am also a Delta Airlines man and have been my whole adult life. Brand names take on a personal aspect within our lives because we align ourselves to the images portrayed in the advertisements, especially the people using the products on our television screens. We ultimately invest our emotions into the brands, and in a large sense they complete us as individuals, just as our clothing, jewelry, or other accessories do.

There is much to love with regard to the modern industrial life, and generally speaking, because of that modernization, I have lived a comfortable life. I enjoy a good stereo system, refined clothes to keep me warm, and running shoes that allow me to spread my proverbial wings on my daily jogs. I enjoy the personal freedom of cars and airplanes, which allow me to visit far-off places. However, there can be a downside to our industrial lifestyle.

Modern industries are adept at translating our wanton desires into perceived needs, and these needs are very different from biological needs. For instance, we all need to eat food, but we don't need to eat the foods that are made by factories; instead we are actually supposed to eat natural foods for optimal nourishment. This fact is mostly forgotten as we have welcomed technology to create and manufacture our foods, just as we have welcomed innovations into every other aspect of our lives.

Having grown up with a fascination for the biological world combined with a prominent American idealism that technology, especially science, is wonderful and powerful, I chose medicine as a career. Therefore, as a physician, my industrial identity took on new potency. Being a physician meant that I was formed by the premier industry in the world — the American medical industry. Although trained as a physician, I was still paradoxically at risk of becoming unhealthy; therefore I am compelled to take the reader through my education to see how my thinking evolved over time. My medical training did not protect me from the detrimental effects of manufactured foods and other comforts, but my unwavering fascination with the ideal form, synergy and synchronicity of the human body did bring me to better health.

A LIFE OF SCIENCE

I have always been fascinated with the innovative products of industry. Science, which is of course generally ascribed to industrial might, interested me as far back as the second grade, when I would read my Aunt Kit's medical encyclopedias, with their overlays of the different organ systems for the human body, and developed an intellectual respect for human form and function. During these early years I also frequented the Museum of Science and Energy in Oak Ridge, Tennessee, where I admired the skills that created the first atomic weapons. Later, my fascination turned to space exploration as my uncle Joe Cain worked for NASA as a physicist and provided me detailed books from NASA about the Apollo space program. Looking back at my identity as an American, I always felt privileged to be from the country that led the industrial world. in my mind, places that were considered "developed" were indeed developed as a result of American influence and in accordance with American principles.

My formal introduction to biology and the cellular aspects of life was in the tenth grade, where Ms. Mote inspired me with interesting lectures that appeared to have great relevance to my daily life. Ms. Mote, who was in her early thirties, had shoulder-length black hair, dark eyes, and a big smile that confirmed her spirited demeanor. She loved her subject matter, and I strived to be her best pupil. Her class was also my introduction to the use of biofeedback. By virtue of the tests the students performed on each other, I saw firsthand how much influence the mind has over the body. I have been intrigued by the mind-body relationship ever since, and much of what I try to accomplish while treating my patients is to promote their mindfulness of their health to effect positive changes in their lives. In fact, I have come to believe that many of our modern health problems are a result of the loss of focus on what is truly important to living a healthy life.

During my undergraduate studies at Florida International University, I was mesmerized by Dr. Landrum and Dr. Makemson, who taught organic chemistry and biochemistry respectively. These scientists instilled in me a respect for the unseen microscopic essence of all life. The biochemical world is indeed wonderful and awe inspiring. I realized that the transfer of energy to sustain life hangs in the balance of electron clouds and the ability of enzymatic reactions to change the probabilities that sustain life from impossible to probable and efficient.

I engaged in my undergraduate studies while working part-time as an Emergency Medical Technician in the Emergency Room at Kendall Regional Medical Center. It was there I first saw the application of biologic sciences directly into human affairs. I experienced the excitement of medical rescues and these interventions being applied to save the person. Each medical rescue is indeed an obvious example of scientific knowledge being applied to improve one's immediate health. These experiences solidified my desire to become a physician.

In medical school, I enjoyed the personal challenge of absorbing the expanse of medical knowledge delivered to me each day. My lecturers were superb, and their expertise in their subject matter was obvious from the minute they opened their mouths. I was in awe of classes with names like Mechanisms of Disease, and I felt I had arrived at a place where I could truly learn about the workings of the human body. On many levels medical school was good for me, teaching me science and the tools to care for people. My medical education also disciplined my work ethic and my mind.

After four years of hard work, I was proud — and I even felt a little powerful — to be part of the class of 1993 at the University of Miami School of Medicine. We were young physicians soon to be establishing our careers at the turn of the century, a time when there was so much promise for new treatments for the old diseases. My wife, Claudia, also completed her post-graduate education in Occupational Therapy, and we moved to Salt Lake City, UT, where the first of our three children was born while I finished my post-graduate medical training in internal medicine at LDS Hospital in Salt Lake City.

As a medical resident at LDS Hospital, I was fortunate to engage intellectually with eloquent and capable physician-scientists such as G. Michael Vincent (who, I was sure, could see into my soul with his piercing blue eyes), C. Gregory Elliott, Corwin Q. Edwards, and Jeffery Anderson. They inspired me even as they demanded excellence from me. LDS Hospital was the cardiac treatment capital of the Intermountain West and was also the major referral center for trauma care and other medical subspecialties. This was an exciting, technology-rich facility, where advances in medical science were being formulated and applied readily. My mentors, although esteemed authors of recent *New England Journal of Medicine* articles, were always available to me for office chats. I had the chance to interact with elite representing all of the major medical disciplines. My three years there were most excellent and prepared me for the career I wanted.

As is the case for anyone in an internship and residency, I was trained by the medical industry and so my life story became entwined with it. It was in the Coronary Care Unit where my internship was initiated, and it was in this context that I developed a more personal relationship with the practice of medicine as well as with the medical industry, which must be included in any discussion of an industrial identity, but especially for a physician. My post-graduate training began with aggressive medical interventions in the Coronary Care Unit (CCU), where heart attack victims are treated and heart failure managed. Excitement is the norm in the CCU because heart rhythms can change abruptly and someone is always having urgent chest pain or changes in their vital signs (clots within the coronary arteries induce a heart attack and subsequent heart damage since the blood flow is suddenly cut off, and this is the number one cause of death in America as well as the cause of a great deal of disability and suffering). The pressures and excitement are part of the ambience as immediate mortal risk charges the air.

Many of the procedures done in the CCU, although technically very sophisticated, are a lot like plumbing interventions in the sense that clogged coronary arteries are essentially clogged pipes. Blood is delivered to the heart muscle by the coronary arteries so, if they become suddenly blocked as in a heart attack, the heart muscle may die (infarction). During a heart attack, these arteries must be opened before death of the heart cells takes place. As with household plumbing, these pipes (coronary arteries) can be similarly opened up with drugs such as T-PA or Rheopro. More procedurally sophisticated approaches to alleviate coronary blockages include angioplasty and stents, which displace the arterial plaque and open the clogged arteries directly.

If one needs a whole new set of pipes because the above remedies are not sufficient, then the pipes are replaced through coronary artery-bypass grafting. But unlike pipe replacements, the new coronary artery grafts are an inferior product to the original set of pipes. In fact, even metallic stents, which remain with the patient the rest of his life, are an inferior product compared to the original un-stented heart. Unfortunately as well, clot-busting drugs carry significant risk of use while only stabilizing the situation until the patient can have cardiac stents implanted or by-pass surgery at a later date. Making matters worse is the fact that, once you have entered the CCU and undergone one or more of these procedures, the clock now ticks toward the next coronary event, a sure thing if ever there was one. Thus, entry into the CCU christens one's relationship with the hospital, an intense relationship that will continue for the rest of one's life.

CCU patients have much in common: high blood pressure, obesity, high cholesterol, and either smoking or drinking too much alcohol. Age used to be an important factor but is becoming less so because younger people are now afflicted with this coronary condition. Family history has always been important, but rather than coronary genetics, I speculate that this is the case because family members tend to have similar lifestyle patterns and tend to view health in a similar fashion. Approximately 15% of heart disease patients are considered to have no identifiable risk factor aside from the fact that they live in America, implicating a potential environmental cause. Asiatic people and other ethnic groups are much less prone to heart disease but increase their risk by living in America. Consequently, early on in my stint in the CCU, I began to get an inkling that there is something important to living in America that is not conducive to good heart health.

As I was finishing my residency training, I began to contemplate which business style of medicine I would practice. Like many others, I considered my private practice dreams, but was business-savvy enough to know that the overhead costs for internists in private practice were soaring in the 1990s. I was not keen on undergoing a business education because I was far more interested in practicing the art and science of medicine. So, upon completing three years of post-graduate training in Salt Lake City, I joined a large and successful multi-specialty clinic in Layton, UT. I continued to participate in medical rescues as much as possible, and our group of internists took calls every night for the Emergency Room at the local hospital. My partners and I performed numerous medical interventions, tended to the sickest patients, performed cardiac stress tests, and admitted most adult patients within the community to the hospital.

I found my practice at the clinic intellectually gratifying, and I grew to love the power of modern medicine, and of course my personal identity became very much aligned with the medical industry. I felt very productive and at home intellectually within the medical community. Indeed, I had adopted a very complex industrial identity and saw good things coming from it. It took me forty years and a great deal of open-mindedness to realize my industrial identity and the overall harm it represents for my health. The impetus for this book came from the fact that I became unhealthy as a middle-aged adult and was required to reevaluate many aspects of my life, especially what I was consuming and why I was consuming those things.

Civilized Life And Comfort

By 2004, I had built a very successful practice of internal medicine, and I was of course enjoying the accouterments of a civilized life. Claudia and I had constructed our dream house, an executive home built of brick and stone, so we were quite insulated from the natural world in Fruit Heights, UT, and we loved our community. We had an active social life, and our three children seemed to have an endless stream of friends exhibiting an insatiable desire for treats. For me, these were all signs of a successful life, and I became quite complacent.

Looking back at the first forty years of my life, I will say that my only activity directly connected to nature was intermittent periods of routine exercise. I did not eat much natural food, however. Similar in many respects to my childhood, my diet consisted almost solely of heavily processed manufactured foods. The only natural whole foods I routinely ingested were servings of chicken or steak, although given what I know now, these meat dishes were less natural than I assumed. Breakfast has always been a difficult habit since I became a professional, and I thought eating on the run or missing lunch altogether was just part of an efficient work ethic. I could see more patients in less time. My life progressively became unbalanced, and my weight ballooned to 210 pounds, fully 55 pounds over my ideal weight.

After several years of practicing medicine, years of treating people with coronary issues that I knew stemmed from shared lifestyle traits, I was very successful professionally but increasingly less so physically. I only vaguely resembled the image of what a healthy human specimen is supposed to look like, which I clearly remembered from my anatomy books and the Discovery Channel shows about our ancestors hunting and gathering. I realized that I was but a semblance of a *Homo sapiens*. My health and self-maintenance had become a distant rumor while comfort foods like brownies, pizza, and ice cream slowly became regular fixtures in the house. Life was good and everything seemed to be going according to our life plans. We had much to celebrate, and we had nights of debauchery called "Lose It" nights, which typically consisted of candy, brownies, ice cream, and pizza while we watched a movie on our new big screen TV. Our lives were busy and there seemed to be little time for self-care.

In the clinic, my patients often complained to me about their muscle aches and joint pains, which they attributed to their age, even at relatively young ages. All of my patients seemed to feel older than their stated age, and I realized that I was beginning to feel the same way. In middle age, for the first time, I felt mortal. I suppose part of my lack of insight was that, by American standards, I was still relatively healthy, especially compared to my patients and some of my family and friends. Acquaintances also said things like, "You carry your weight well." And I still did not have high blood pressure or high cholesterol. I would be incredibly sore, feeling physically abused even, the next day, but I was still capable of jogging 1-2 miles (albeit slowly), perform strenuous yard work, and even water-ski or snowboard. I still maintained the courage to preach the virtues of a healthy lifestyle, like the importance of diet and exercise, to my patients. In short, a different image of me than what I actually was still resided in my mind. I was not like *them*, the infirm, the unhealthy.

Eventually, I did not feel comfortable in my own skin, however. Deep down I knew my message as a peddler of health-care was corrupted, filled with hypocrisy. I considered myself a successful American, but that overall image of success did not include a person, an organism if you will, successful at taking care of himself. I became more self-conscious and my self-esteem lessened, which was scary indeed because, as a comfortable fat cat (all puns intended) living the American dream and much too content to pay attention to my health, I was not alone. I was obviously suffering from an unbalanced life and yet felt trapped in it, and I realized that most people I knew suffered from the same condition.

Having now reached a weight well over 200 pounds while feeling all the more "older" and more limited, it became obvious that life changes were needed, but I was not sure what they might be or how to achieve them. In the mirror I saw a rounded face, a chest that once had defined pectorals now closer to swollen breasts, and an abdomen that did not have any muscle definition whatsoever. My belly did not protrude terribly, but only because I could thrust it inward to even it up with my chest, and I will also admit I had dimpled buttocks. I was not at all comfortable, but some days the image of what I had become would be shocking and other days somehow curiously tolerable.

I still find it incredibly interesting that, despite all of the unhealthy meals and a sedentary life, I did not expect to look like this. Stranger still, despite my medical education and my knowledge of what constitutes human health, I still needed to suffer from obesity. It is also a great curiosity that the shock I felt some days when looking in the mirror did not provoke me to action sooner. How was it that I came to look like this, and moreover, once I realized I had fallen to such a state, why didn't I act?

At thirty-nine years of age, my pants were getting tight, and I began to feel frequent aches after strenuous yard work. I adapted to my poor physical condition by hiring a lawn service and buying larger pants. With flexible waist-banded pants, I could buy 34-inch-waist pants and feel comfortable all day even though I deserved a 36-inch waist. In fact, I distinctly remember the joy I felt when I discovered my preferred brand of dress slacks at Sam's Club, now with elastic expandable bands in the waist. Hectic workdays required walking up and down the clinic halls, and I was called to the hospital often, and so for comfort I adapted by purchasing orthotic in-soles for my shoes.

Of course, these unfavorable physical changes exact a toll on one's physical performance and can change one's life experiences. My performance noticeably diminished with my added weight, as evidenced by loss of speed and power. When skate-boarding with my kids, the extra weight threw my balance off; and my excursions into nature diminished to an occasional vacation in Moab or an infrequent family picnic. Worst of all for me, I became less capable of summiting the magnificent Wasatch Mountains that were always in my sights because my muscles had atrophied and were concealed in layers of adipose tissue.

My identity as a unique biological creature had changed. It happened without any recognition on my part, and it did not happen overnight. I am not alone in this experience. Metaphorically, we are born as panthers or some other wild beast, born to be hunters, but after years of sloth and gluttony, we have turned into fat cats, loungers extraordinaire. I have spoken of my acquisition of an industrial identity as manifested by the brands I have been loyal to in my life, but our industrial identity, as it relates to the foods we consume, manifests at a cellular level, and the change is not a positive one. The consumption of machine-made products also promotes a transformation of our personal identity with regard to our physical being and sense of self. Our industrial identity is in fact an alternate identity that is created and maintained by industries, therefore a divergence from our more natural identity that would be realized if industries were not such a large part of our lives. Recognition of our industrial identity causes us to question many existential matters in life such as the health we sustain within this manufactured identity.

Most Americans do not place an emphasis on health as a measure of their success. I will also say that most Americans do not consider health to be of supreme value but more as a given, a fact that is determined by forces greater than they, and as a result few people engage in routine exercise and a diet consistent with optimal health. Instead, we measure success within the context of, indeed in terms of, our industrial identity. Consequently, most Americans are stricken with chronic diseases and/or obesity, and our society is burdened by the physical, emotional, and economic cost of disease; and yet most Americans consider themselves as extremely successful people. The irony is nearly overwhelming.

These Americans, sickened by obesity and chronic diseases, also believe they exercise an inordinate amount of personal freedom. Presumably most people will say they do not want their disease or their obesity; then perhaps we are not so free as we assume and something is fashioning our lives and our individual destinies that is outside our control. How else do we explain creatures with free will ending up in a state they do not choose?

I will argue that we are in fact trapped into an industrial-made alternate identity that is contrary to our biological identity, our identity as *Homo sapiens*. I will argue that our methods for living, our expectations for life, and even our fundamental perceptions about life become altered as we consume industrial goods. We have transitioned from a biologically inclined life to an artificially-based life with the infusion of manufactured products, and there appears to be a dose effect as well: the extent of our biological corruption is proportional to our level of consumption of industrial products. Much of our society is developed from the products of industry, not solely the material goods, but also a set of tenets by which we live that are put into effect by the industries making the goods we consume.

THE INDUSTRIAL ORDER

Industrial Order: 1) the collective force of industry operating on mankind; 2) the culture of dependence created and sustained by industrial products; 3) the universal application of industrial products to provide for man's physical and emotional security; 4) the progressive evolution of technology that increasingly corrupts natural forces and therefore competes with biological forces; 5) the prevailing industrial scheme that entices and coerces man to abandon his most basic instincts, thereby changing his destiny and his essence.

My industrial identity, formed long ago by constant exposure to ads for products to consume that I was told were fun and meaningful in some extraordinary fashion, sexy and filled with some unspoken potential and therefore good to incorporate into my notion of my own being, led me to a terrible state. In fact, by the time I was forty years old, my biological identity had been altered, surreptitiously and insidiously — and, as measured by the previously mentioned data, this is not only happening to most Americans but beginning to do so at a younger age. This is not only my perception as someone who has realized his industrial identity and the toll it has taken on his health but as a physician, and not just as healer but as a member of the medical industry.

Americans are struggling with chronic illnesses and physical disabilities that are not necessarily related to aging. In fact, the young and middle-aged people of America are developing chronic diseases that were seen only in very aged people a couple of decades ago. Young people, who should be those among us who have the most health potential, are becoming not only chronically sick, but twenty percent is obese and one in ten is on daily prescription medications. One-third of all US armed forces applicants are rejected due to physical dysfunction. Hence, our identity as a nation is changing, but who or what is controlling the change?

This unhealthy, and indeed sad, state of affairs is occurring despite the overwhelming application of increasing technological wizardry and the inordinate expenditure of national resources. In truth, the issue is not technologically resolvable or a function of a lack of financial input. Average Americans merely suffer from a disconnect between their beliefs about health and their biological being, the innate circumstances of being an organism. This is at the least unhealthy thinking and perhaps amounts to insanity. How is it that we lose touch with our biology, giving up attention to the natural processes that give us life, if not a kind of madness? The disconnect appears to be innate to our industrialized culture, but human beings also seem to be pretty gullible, rarely questioning the values of their culture, and both these concerns will need to be addressed if we wish to be a vital people.

American culture is a prime example of the Industrial Order wherein our whimsical desires are translated into perceived needs for technological wizardry. As I will point out more clearly later, technological evolution is misinterpreted as human evolutionary progress, as in improvements of the human form. A fantastic variety of consumer choices creates an illusion of personal control and personal freedom, and yet, consumer reactions are quite predictable, which suggests that these reactions are programmed. There would be little success in the advertising world if we could not be made into predictable consumers. We are influenced, coaxed, and coerced into actions, even to our detriment, that benefit a given industry. Corporations shell out huge sums of money on consumer research so they can utilize our tendency toward malleability to increase the company's bottom line.

The food and medicine industries have worked synergistically, and one could even say interdependently, to change our individual identities as well as the identity of our nation and our overall notion of what it is to be a human being. The consumption of manufactured foods creates the need for manufactured medicines, medical tests, and medical procedures. Indeed, active consumers of manufactured foods, which is most Americans, ultimately become industrial inputs for our prestigious medical centers. The predictability of this outcome and the relationship between the two industries does not require much imagination or scientific knowledge to forecast or assess. This is a simple observation that anyone can make.

The local donut shop with its conveyor belt of donuts (among many other processed food companies) is consistently manufacturing the raw materials (donuts) that eventually corrupt our physiology in such a way that our health changes and therefore we change as people. The Industrial Order becomes enmeshed into our body's fabric as we chew on the Krispy Kreme donut that just came off the assembly line. Indeed, like trained lab rats, we know we are guaranteed a freshly-made warm donut when the red light on the store front is flashing, and so we head inside, buy them by the dozen because they are cheaper that way and everybody loves a deal. What we fail to understand is that these donuts create the need for coronary procedures and stroke centers. With incredible predictability, Americans who routinely visit bakery shops, candy stores, fast food restaurants, and all-you-can eat buffets will also need to routinely visit the portals of entry into the medical industry as sick patients.

As an aside, I have already suggested that the food and beverage industries have changed our psychology and our biology, but they have changed one more thing by virtue of these other changes: our fates. Many of my readers will know of someone who has the condition of sleep apnea, a condition usually associated with an overweight person who cannot breathe adequately at night to keep oxygen levels normal. These people chronically suffocate as a result of their body size. While sleeping, they are drowning in their own fat and skin, but relief is provided if they can tolerate the application of a mini-respirator. These are portable contraptions that apply positive pressure to blow air through the excess, redundant tissue between their mouths and their lungs. These people are dependent, first on manufactured foods and then on the respirator, and thereby not only do they become dependent on the manufacturer of these devices but their fate changes: they are now people whose sleep is interrupted, who must be connected to this machine, who must live a particular way because of their industrial identity.

In a very similar fashion, the donuts, bagels, cereals, and other manufactured delights will cause dependence on the pacemaker industry to keep one's heart beating, another technological intervention that is becoming quite common. People who subsist on processed foods are also more likely to experience kidney failure as they develop hypertension and diabetes throughout their lives. They will eventually require dialysis machines and the medical centers where those are available. The application of dialysis is a multi-billion-dollar growth industry in and of itself.

As I have proposed in my definition of the Industrial Order, technology breeds more technology, and the aforementioned industrial relationship between food and medicine illustrates the concept quite well. Incredibly, the human host that provides the medium for these technological gadgets to procreate is usually oblivious to such matters.

BROKERED EXPECTATIONS

As a modern physician, I am an intermediary between the patient and the complex world of modern medicine. Indeed, an extremely intimate relationship develops between a patient and the medicine industry. Under my care, the distinctions between the artificial, the manufactured, and the natural world become blurred as the patient accepts microscopic and macroscopic artifacts such as medications and devices enmeshed into his body. There is an obvious competition between the manmade interventions that I can offer and the patient's natural ability to care for himself. Patients do not seem to consider their natural power inherent to life as being equal to (let alone better than) the technological miracles of medicine. In fact, I have sensed a tension, a push-back from patients, when I try to steer them away from simply opting for medications to treat their symptoms and toward methods to treat their underlying condition. Medical innovations tend to win us over by the perceived power and appeal of modern technology, but then we already have that notion embedded in us by the time we move from industrial food to industrial medicine by virtue of having achieved an industrial identity.

Complements of the Industrial Order, people tend to obtain their expectations for health and medical care from media outlets and their friends. For example, in some recent pharmaceutical advertisements, people are told that all of their disabilities just happen to them (implying mysteriously) as a matter of a normal life course. This is important to recognize because it instills the notion that the incorporation of the artificial, the newest medical innovations, into their physical being is simply the next logical step. The ideal dream for the medical-industrial complex is to make people believe that the transfer of their lives to artificial life support systems (obviously there are varying degrees of this) is a natural progression. Unfortunately, the comfort gained by the application of these devices is not equal to the health that is lost by eating manufactured foods or the level of health possible outside of the Industrial Order.

People do not expect exercise and a sane diet to be as valuable or powerful as medical science, but this is a fallacy. Even interventions of the artificial variety require some aspect of good health in the natural sense in order to function properly. It is remarkable the number of patients who do not understand that powerful antibiotics only work if their immune systems are working properly. Their bodies must be strong enough to fight infection, or the antibiotics simply do not work. In short, artificial interventions are only working upon the natural foundation of our health anyway, and so the assumption that natural interventions will not be successful is a logical fallacy.

However, the basis for this misunderstanding is the medical-industrial complex itself, both in its self-portrayal relative to health and in our powerful industrial identities, not to mention the gee-whiz nature of much of our latest medical technology. Working in medical centers, I routinely see that a patient's potential for healing himself is discouraged. Every day I witness an individual's power over his health outright usurped by the medical industry. It seems innocuous but is truly insidious. People are entranced by the physician's fancy tools, professional demeanor, and credentials, including his education and government license. And of course the medical professional, the medical industrial identity a very large part of his overall industrial identity, will tend to push those artificial interventions above more natural remedies. The patient almost always buys in, seeing both utter expertise and the hope of new technology as his health salvation.

Moreover, this machine, this industrial complex that is modern medicine, has become enormous. Not only has the amount of expertise and technological sensationalism gone up exponentially with the advent of specialists and the ever growing number of devices available to adjust the patient's biochemical parameters according to the latest evidence-based guideline, but the patient is now moved through this machine as if on a conveyor belt so the personal touch is all but gone.

Powerful specialty groups and organizations are constantly developing additional protocols, systems, and entire wings of institutions to treat specific conditions, and specialization is touted to guarantee a better standard of care applied uniformly to all patients. On first glance, most people would say standards of care could only be beneficial, that standards and so many professional minds at work on a given body should amount to superior health outcomes; but somehow the benefits to society don't seem to materialize as in theory. Pamela Hartzband, MD, and Jerome Groopman, MD, echo my concerns in the October 13, 2011 *New England Journal of Medicine*. They admonish the medical industry because it contends that "clinical care should essentially be a matter of following operating manuals containing preset guidelines, like factory blueprints, written by experts." A patient's symptoms can only be addressed on a more personal level after all of the conditions of the machine or the conglomerate have been met. Standards of care represent the further mechanization of patient care.

Factory quality measures (standards) are good for assuring that widgets are fabricated in the same manner, as in perfected inanimate forms, but people are very different, not only from widgets but from each other. Most people do not understand the reasons for and nuances of standards of care and therefore they are apt to deviate from them. Pressure is then placed on the medical systems to improve patient "compliance" — as in industrial compliance.

Physicians and medical systems are judged by these standards of care, so the pressures to comply are enormous, usually enhanced by money. Front and center, the doctor-patient relationship has suffered because the nature of the relationship has changed. Personal touch has been replaced by factory-style efficiencies. To assure the standards are kept, teams of medical professionals, including nurses, information experts, case managers, discharge planners, and even specialized Medicare compliance officers, are incorporated into the routine medical transactions while corporate propaganda is disseminated, making sure everyone is on-board. Somewhere in this maze is an individual with a need.

Adding to the confusion within the context of a patient's concerns is the fact that many healthcare professionals tend to exhibit ownership of patients' medical conditions, perhaps as a result of their training to intervene and their sense of responsibility, but also because they are judged on patient outcomes. Deviations from the standards are not tolerated even if they are the patient's fault, and this strengthens the compulsion towards the medical system's ownership of a patient's illness. However, such ownership by third parties will usually sever the patient's relationship with his own condition.

Patients find it difficult to secure their role in the medicalized assembly line. I cannot help smiling when I hear members of our medical team express frustration when a patient's outcome measures do not fulfill the medical team's desires, and quite often it is deemed the patient's fault. "They (the patient) don't get it," is exclaimed, but how much information is there "to get" for a patient who walks around with multi-organ failure, a common problem in the general population these days? The only aspect of multi-organ failure a patient should understand is to avoid it.

Even if a doctor is inclined to suggest more natural remedies outside the medical-industrial complex, with an industrial identity firmly established in their lives, Americans expect to have health manufactured for them. Indeed, modern medicine has provided treatments for many scourges, but with that success comes the eventual compulsion to apply modern technology to all of our symptoms, without pausing to consider the source of our physical maladies or more appropriate natural remedies. Patients and doctors alike are consistently trained through the Industrial Order to relieve symptoms, such as a headache, with pills or procedures. One of the last things a person who walks into my office with a headache considers is his deficiency of self-care, an idea that appears to escape the minds of most of my patients completely prior to their visits. Quite often I recommend that my patients should drink water to rehydrate their depleted bodies or take a much needed nap (Americans are notoriously sleep-deprived), but these simple remedies are overshadowed by the consideration of a CT scan or MRI of the brain, "just in case." In my experience, tests never remedy headaches and self-care generally does.

The most interesting aspect of this entire scenario is the fact that the medical empire simply reflects our desires. Americans want their "health" to be a product of industry, just as all other aspects of their lives are a product of industry. Health should be manufactured, then certified, then bar-coded, and finally mass-consumed. The ease with which we accept industrial products into our lives lowers the threshold for acceptance of medications and medical procedures. In fact, not so long ago, most people were reluctant to ingest industrial-made products, especially not wanting to consider themselves dependent upon them. Now, however, patients almost demand some pharmacological intervention just as they might demand that a technician fix their cell phones. Therefore dealing with unrealistic expectations is a large part of my job, unless I choose the easier route and stick to the standards of care.

Sometimes I ask people why they trust industries enough to consume potent products on a daily basis, and they will usually give me a blank stare. They had never thought of why . . . it just happens. But did the industry in question do anything that has garnered such trust on such a personal level? Many people stipulate that we should trust modern medical institutes because the government is watching them, but they have forgotten that the government *still* subsidizes tobacco production.

FAITH WITH RISK

For at least the first decade of my career in medicine, I was convinced that the medical industry always had the patient's best interest at heart and espoused the most optimal methods to secure the best health possible. My patients invited me into their lives and shared their ups and their downs and by doing so taught me much about life. I also enjoyed the detective aspect of being an internist. Every patient check-in slip is potentially a mystery I am asked to solve. My job as an internist is to make sense of a large number of personal complaints in the context of lab tests, to determine the nature of the patient's malady, and to contemplate potential remedies. But as my industrial identity had taken on some ominous physical characteristics, I have since learned to question the tenets of society that caused, or at least allowed, me to acquire those same physical characteristics of my patients.

I believe there is a prevailing faith in the medical system that often translates into an existential human threat. The most serious threat is our failure to realize the health we can achieve through how we live our lives (this will be explained more fully later). An additional threat, which is separate from and yet magnifies the original threat, is to put our hope for health in the hands of an industry. I distinctly remember the day when I felt for the first time that the tools of medicine may not have been created with the patient's best interest in mind. I was approached by several pharmaceutical representatives during any normal day at the office, each one's primary purpose to influence my choice of medications to prescribe to my patients. The day my naiveté ended was a typical clinic afternoon.

A pharmaceutical representative named Amber showed up at my office to discuss the treatment of the metabolic syndrome using Avandia. I really liked Amber because she was one of the most professional of the drug representatives I routinely encountered. These reps always come sharply dressed with bright smiles and cheesy tokens or gifts (large gifts are outlawed unless we are paid as "consultants"), but they are not all the same. Amber was one of the few I ever befriended (for the most part I instructed my nurses to dismiss visiting pharmaceutical representatives quickly).

Avandia was originally developed for the treatment of diabetes, but its cost was much higher (as is the case for all new drugs) while its relative efficacy for treating diabetes was much lower compared to existing generic medications that were much cheaper. The side effects of Avandia were also less desirable and included weight gain. Avandia obviously had hurdles in its attempt to gain market share for the treatment of diabetes. I would only consider using Avandia as a last resort, when all other measures have failed. GlaxoSmithKline, the manufacturer of Avandia, also sensed these barriers to obtaining its share in the diabetes market. As expected, Avandia's entry into the marketplace for diabetes treatment was not satisfactory for the corporation's return on investment, and so another use for the drug was found. This is not unusual.

At the time, articles were beginning to appear in medical journals describing what was termed the "metabolic syndrome." Patients with metabolic syndrome are typically obese and sedentary, meaning that the best treatment would be to fix lifestyle issues. Scientifically, and as a medical professional, I saw little value to diagnosing any of my patients with metabolic syndrome since the treatment should be to improve their lifestyle. In fact, to this day, more than ten years later, health professionals still debate the utility of classifying patients with metabolic syndrome, a condition that, with its growing prevalence, is slowly becoming a normal state of American living.

This particular day, Amber entered my office to formally discuss the metabolic syndrome for the first time. She was very prepared, but fortunately, I was prepared as well, and our discussion was lively. Amber also brought sample boxes of Avandia, even though she knew that I was not a fan of Avandia for the treatment of diabetes, but she was also promoting its use for the new condition, the metabolic syndrome. Amber pushed me to consider using the diagnosis of this new condition more because the research she brought with her showed a benefit for the patients. I was not convinced, but this type of interaction was not unusual.

What really caught my attention was the fact that Amber also wrote down the current procedural terminology (CPT) code created just for the metabolic syndrome, the condition that still was not fully agreed upon in the medical literature. We use CPT codes to describe our patient's condition in order to justify procedures and medications to the insurance companies. That is, we obtain our compensation for services by way of the CPT code. As put forth by the American Medical Association, with this new CPT code I could bill the insurance companies for my services that addressed the metabolic syndrome and the patients would be able to use their insurance benefits to purchase Avandia. There was no other drug available that was deemed "indicated" by the Food and Drug Administration (FDA) for the treatment of the metabolic syndrome. Therefore, insurance companies were required to cover the cost of Avandia. My eyes were opened wide when I saw the sudden certainty in a hotly debated syndrome, a new CPT code for it, and a medication for the condition that was originally designed, developed, and tested for a different condition.

This was a very succinct presentation of the way medical business is performed. Amber's strategy, which of course her company instructed her to use, was extremely well planned and transactionally complete for all parties concerned. The doctor, the pharmaceutical company, and the insurance company were seemingly united for the benefit of the patients, who are expected to be happy because they are getting the latest drug available. As long as the patients do not question this version of health, we can all be satisfied with ourselves and make a little money in the process. To this day, I still maintain there is no value, no improvement in patient outcomes in diagnosing the metabolic syndrome. Patients can treat obesity, high sugars, high cholesterol, and hypertension, the elements of this syndrome, by lifestyle modifications, and I am not convinced that medications will improve or save anyone's life.

Avandia went on to be a billion-dollar-plus per year revenue generator for the company, but of late it has come under fire for substantially increasing the patient's risk of strokes and heart attacks. The FDA now has a black box warning and special protocols to be followed if a physician is using the medication for the treatment of diabetes, a condition more serious than metabolic syndrome. As of November 2011, GlaxoSmithKline has been ordered to pay $3 billion, the largest drug settlement in history, to settle United States government civil and criminal allegations regarding its promotional practices of Avandia. The law suit stipulated that GlaxoSmithKline manipulated its drug research and then paid doctors to prescribe the drug despite the suppressed evidence that the drug was not only ineffectual but causing thousands of strokes and heart attacks.

Since that day in my office when I was introduced to Avandia and the metabolic syndrome, I have become more conscious of the overt industrial forces within the practice of medicine, and I have become much more skeptical of the medical industry in general. GSK is not the only large pharmaceutical company to be found corrupt, as in misrepresenting medical data or bribing physicians for positive results; virtually all of the major companies are also guilty, including Merck (Vioxx), Johnson and Johnson (Risperidal, Natrcor, Invega), G.D. Searle (Bextra), Pfizer (Celebrex), Eli Lilly (Zyprexa), Abbot (Depakote), Novartis (Trileptal, Zelnorm, Sandostatin, Tekturna), Astra Zeneca (Seroquel) and Takeda (Actos). Perhaps some of the most egregious harms done to patients have been with the implantable medical devices such as heart stents (Guidant), pacemakers (Boston Scientific), and artificial hips and knees. The FDA does not evaluate the safety and effectiveness of medical devices as rigorously as people are led to believe. An elegant ancestry of the flawed metal-on-metal hip implants was recently delineated in the January 10, 2013 issue of the *New England Journal of Medicine*, wherein the authors of the article showed that the FDA cleared the DePuy ASR XL femoral heads despite the lack of safety and effectiveness. These innovations are being incorporated into our bodies and our lives.

I personally know that the power of modern medicine is immense and that it can be an honorable profession, but the encounter regarding the promotion of Avandia mentioned above cracked the illusion that the medical-industrial complex always has the patient's best interest at heart. Acquiring new customers (chronic industrial inputs) frequently overrides safety concerns, efficacy and common sense. For an individual, it is one matter to put our trust in an industry when we have no choice in life, such as in a critical moment requiring emergent surgery, but it is another matter to trust an industry with our long-term health prospects, especially when we have a great deal of health to lose by taking the wrong path. A prudent question to keep in mind as we interact with health care providers would be: are they providing for our best health or is there a better way?

A CULTURE OF RESCUES

The American desire for chronic rescues as a means to health lacks sincerity for our lives. This is typical of the Industrial Order wherein our desires are made into needs. We would not need medical rescues if we would live with more sincerity and adopt more natural means for health, like avoiding manufactured foods. Our desire for the sedentary life, also comforted by the consumption of manufactured foods, creates the desire for medical rescues.

In his masterpiece, *Consilience*, E.O. Wilson provides a definition of culture put forth by Kroeber and Kluckhohn: "Culture is a product; is historical; includes ideas, patterns and values; is learned; is based upon symbols; and is an abstraction from behavior and the products of behavior." In short, people create the culture in which they live, but the culture also shapes the people born into that culture (it is learned). In modern American culture, the individual members of this society appear to be on autopilot in this land of plenty that has provided endless inventions, not only never questioning our industrial culture but developing an insidious, deep-seated predisposition within us — a relationship to and dependence upon both industry and technology for life.

We trust our technicians and medical scientists, largely because they provide us products that appeal to our desire for comfort. Consequently, very often we impetuously apply medical technology with little concern for side effects or holistic soundness. The power of scientific reasoning prevails in the micromanagement of the world, but this removes our attention from the larger picture, from the wholeness of our physical nature. Arguably, common sense and intuition as well as science may be important in our health care, but these ancient instincts are considered relics in the face of our society's "enlightenment" through our acquired knowledge, especially that gained via the scientific method.

Medical rescues can be dramatic, or they can be more of an ongoing nature, such as taking cholesterol and diabetes medicines to counteract the dietary excesses of fat and sugar respectively. This practice of chronic rescue has entered our culture unto becoming an expectation. However, although aggressive medical interventions can be incredibly beneficial, only in limited scenarios, such as gallbladder surgery and appendicitis, do they carry the potential to reestablish long-term health. Diseases such as diabetes, hypertension, high cholesterol, arthritis, and heart disease are by definition chronic, and the available industrial interventions do not usually provide significant opportunity for real health, as in health without the condition. Chronic diseases have become an attribute of the American society, seemingly as if there were no other choice but to contract them, which is also a learned cultural element.

Today, there are in fact myriad forms of medical rescue, and many devices and medications that can be classed as chronic rescues are available to the average American and in effect perpetuate the rescue mentality in the American people. This not only causes us to be more prone to a rescue mentality but also scientifically affirms the notion that chronic disease is an anticipated event in our lives, another learned aspect of our industrial culture. In short, the American medical industry has made medical crises part of an average American's cultural expectations and a patterned way of thinking, largely through a feeling of comfort we humans have as we relate to our innovations.

Although we may be comforted by the thought of the technology being there in our time of need, these artificial interventions come with side effects, not only for individuals but for the culture overall. We are weakened as a people the more we become dependent upon medical devices. Each medical rescue will be followed by another down the road, because the people at most risk for needing a future medical rescue are those who have already needed one. Furthermore, when we hear about other people's medical rescues, we are conditioned to seek our own. For the individual this cycle means a reduced quality of life and for the culture a reduced notion of our own humanity. Lost within the excitement of a medical rescue is the thought that how we live our lives is extremely important to the life we have. Lost is the notion that real health requires a natural sincerity for life. Reliance on the medical-industrial complex is promoted heavily in our society, but to what end?

As a business model, the consistent application of medical rescues is an excellent way to pacify a population that resides in a perpetual health crisis. Our membership in the culture of artificial comforts guarantees us a seat in the culture of artificial rescues, a culture which sustains perpetual profits. That is, the widespread consumption of manufactured foods causes a growth in paramedic trucks, and now helicopters, to ferry stroke and heart attack victims, not just those people suffering from trauma. The medical rescue mentality is pervasive and potent. We are tempted by rescues because we as a people feel more sophisticated in a superficial and conceited way. We tend to measure the degree of our society's sophistication by the number of medical rescues we can achieve in a normal day. People are now making decisions about where to reside based on the available access to the medical machines, a different philosophy for a healthy life. Artificial life-support is not only deemed necessary, it is becoming an inspirational element for our existence, an important factor in how we consider our future and ourselves.

Traps Of The Industrial Order

As a physician, a broker of people's expectations for health, I sense numbness and apathy induced by the industrial identity, a particular suffering of the soul that is quite nuanced and yet profound. As if our routine corporeal pains associated with an obese, atrophied body were not enough, there is an existential loss of a natural identity. In fact, our detachment from a more biologically sound vision of existence is quite capable of extinguishing our natural existence altogether. An industrial identity is made infinitely more potent as people enter the realm of modern medicine.

For instance, most Americans do not realize their innate power over personal health, seeing only the power of modern medicine. Science and medical technology tend to trap us into specific modes of thinking that can change how we relate to ourselves and ultimately transform our personal identities. It is ironic that living, biological creatures such as ourselves would perceive industry-based healthcare as the optimal paradigm for establishing health, as in a healthy life. The factory-style efficiencies being adopted by modern healthcare facilities should dissuade us from, not entice us to, them as a source of "health." Shouldn't we know better?

Being a medically dependent society is expected when we are "scientifically" convinced that the industrial healthcare machine is the purveyor of health, as if health required a purveyor. The average person is defenseless against such a scheme because he is not generally equipped to dispute scientific recommendations and therefore is susceptible to the fallacy that health as prescribed by modern medicine is as good as or better than real health secured by more naturally sincere means. Who are we to argue?

Admittedly, natural interventions are not only less sensational than some newly contrived treatment plan but also more difficult to implement after our existence becomes dependent upon artificial treatment plans. Moreover, natural interventions require a sincerity to life that is not always apparent in how Americans live their lives. We need not be so serious if we are living "the dream." The Industrial Order creates the dream and governs our affairs.

Health is not a commodity of interest to many people or worthy of a conscientious effort on the part of a patient (or a person seeking not to be a patient) until it is gone. Even then we fail to see value in our own actions because the culprit in our demise must lie in something else, such as an LDL molecule, a blood pressure greater than 140 mmHg, or perhaps a deficiency in mammograms or colonoscopies. The real value inherent in healthy lifestyles is not sensationalized in a society that relishes complex disease patterns and the intricate medical rescues that must be contrived to attend to such complexities. Our perceptions of and focus on life are fatally biased. The long-term physical and mental effects of medical treatments should not be underestimated. A reliance upon medical rescues tends to trap us into a reduced form of existence.

Technology is generally viewed as empowering even when our innate biological power is being diminished by these innovations. Take, for instance, the notion of an empowered patient. On its face, this appears to be a good thing, and the concept is promoted throughout the media as a valuable way to secure one's health. But is any person really empowered if they remain a patient? The CDC and other distinguished bodies of public health maintain that eighty percent of America's diseases can be avoided by healthier lifestyles (as I will explain in detail), and therefore real empowerment means not being a patient at all. I would argue that the idea of an empowered patient is somewhat oxymoronic, given that very important aspects of our humanity are often lost as we become invested in "health" that is promoted through the powerful innovations and the reductionist methods employed by modern healthcare. The most empowered patients may believe they have increased power over their health when in fact they have merely become more deliberate consumers of the bright and shiny innovations of the Industrial Order, the products that usurp their power in life. Most healthcare consumers remain unaware that they are placed in an artificial bubble, one that is conjured out of the spell of scientific reasoning and man's most potent reductionist methods.

Modern healthcare is premised upon reductionism, the belief that every phenomenon can be explained by analyzing the most basic physical mechanisms in operation during the phenomenon, and also upon the practice of magnifying a mechanism until we lose sight of it. The healthcare industry in fact represents an existential threat to many Americans by virtue of elevating reductionism to the most important method to health when it is at best a rescue, and, at that, it is a Band-Aid when something vastly more far-reaching, albeit far simpler, is needed. Scientific reductionism reduces us into bits of information and therefore very potently obliterates our role in our health affairs.

Technological innovations such as microscopes are made increasingly powerful, and admittedly the view through the eyepiece of the microscope is quite fascinating. We are likely to be convinced that the smaller vision is superior to a larger vision of the same thing, and we are apt to feel power over the process or object being studied by virtue of our ability to peer at it, by virtue of this technological marvel. As we peer into the microscope, it is usually difficult to remember that we have in fact reduced the scope of our vision, mistaking the newest things we can see for the most important. Perhaps our health is not so complex an issue after all, and indeed there is little to be discovered about achieving health by looking at its smallest constituents. Perhaps achieving health is simply a forgotten skill.

Our scientific revolution has advanced by virtue of reductionism, and it continues to drive our innovations and the design of most of the manufactured products we consume, including our manufactured foods. Although reductionism has in essence created our society, including our conveniences and our entertainment devices and much else we deem good, proof of the downside of reductionism is that science now serves to sustain the epidemic of chronic disease we are currently facing rather than eradicating it, no matter the promises or claims to the contrary. Complicity with our unhealthy behaviors is provided rather than true needs being addressed.

Only through a reduced vision can we miss the fact that our entire society, hundreds of millions of people, move through the stepwise process of medicalization — initiated by routine cellular dysfunction, then progressing to the assignment of so-called "pre-conditions" (such as pre-hypertension, pre-cancer and pre-diabetes), then actually accumulating medical diagnoses, and finally proceeding to multi-organ failure — as a giant herd. This very succession has in effect trapped our society by a vast and intricate system of rescues, procedures, medications, devices and facilities. We have become oblivious to the life that could be had otherwise.

The medical arena is an artificial world created by a profitable industry, admittedly the sine qua non of modern man's control over his environment, but more aligned with a desire for customer satisfaction (similar to manufactured foods) than a method to secure a sound biological existence and the most healthy life possible. Under the auspices of reductionism in our healthcare, we are becoming less capable people and therefore less able to adapt as biological people to our industrial-sized problems. We are creating a healthcare nightmare on the horizon in our lives and at the same time building a healthcare empire to trap us and our resources. Optimal healthcare serves the healthcare industry better than it serves our health.

It is important for Americans to recognize that, despite all of the fancy gimmicks and perceived power of modern medicine, the largest explosion of preventable, chronic diseases ever in the history of mankind has occurred as a direct result of modern medicine and scientific reductionism. Modern medicine is not an antidote for the incredible harms caused by the modern food industry, but it is an effective distraction.

Blinded by science, we can escape into the artificial world and our natural identities remain concealed, cloaked within the identity that industries create for us. The scientific method and the reductionism practiced by modern medicine conceal the fact that we are not tending to our own human needs, and therefore a concept of health that is inferior and suboptimal is sold to us as if it were superior and optimal for us. In a word, we are duped by medical experts and their treatment plans that steal our power, pushing us away from personal health and into chronic disease, because they do not consider us whole. Specialists attend to specific diseases and individual organs, each specialist demonstrating fantastic success within his limited scope, perhaps winning a battle or two, but most often losing an entire war because we are not being treated as a whole organism. In order to prevent industrial diseases, we must address our insincerity with regard to a biological life — but what is provided in medical care is more in line with complicity, which compounds our insincerity.

Personal habits that sustain one particular chronic disease tend to cause sickness of all kinds, and therefore the treatment and the management of chronic diseases, no matter how scientifically sound and meticulously managed, will breed more sickness in the person because the crux of the problem is not addressed. Stated the other way around, the medical industry, using premier methods to control disease, often actually promotes disease. Many treatments for chronic conditions generally keep patients sick rather than re-establish a healthy lifestyle, a life that creates health.

When a physician prescribes medications that only mask the vital signs or the measurements pertaining to the patients (such as blood sugars, cholesterol, blood pressures, etc.) but does not endeavor to correct the underlying problem, the patients' unhealthy behaviors, the physician has demonstrated complicity in the patients' real problems, those same behaviors. Once a medication is started, the patients are expected to remain compliant with medical care, and implicitly therefore compliant with the unhealthy pattern of living that caused this problem in the first place. If they change their behaviors, the patients would not be able to acquire the medications and the physician's power would be diminished. And if the patients can get off the medication, then why were they placed on it to begin with? The medical industry's complicity with unhealthy lifestyles is tantamount to treating people as reduced human forms.

It is most interesting that this most devastating predicament is unfolding as the Industrial Order makes us believe that we do have control over our affairs. The truth that we have less control over our health and our lives is concealed from us even as our control is being usurped. What is actually happening is that the medical industry has assumed power over us. Indeed, it seems that reductionism, as delivered by many facets of the Industrial Order, achieves what it aims to do: it reduces us.

The majority of Americans should not look forward to belonging to the medical system, but we are made to feel guilty if we don't follow medical recommendations. To that end we are imbued with a desire for compliance with the latest medical guidelines and gadgets. People are routinely encouraged to be screened for "potential disease" and seek medical care even when they are not sick. Disease awareness days, "health" fairs, and disease-specific health messages are streaming through our consciousness consistently. We are coerced to comply despite the overall data indicating that all of these messages and screenings have done little to improve our population's health (although they do tend to increase the consumption of medical services). Our hope for health is presumed to be a result of the medical care we are to receive, and, I would argue, we are in effect preyed upon. But modern health care is no ordinary predator because it also establishes a parasitic relationship with people. We are actively recruited into the parasitic relationship that affects all Americans either directly or indirectly. We are the hosts needed to sustain the American medical machine.

Successful parasites tend to weaken the host (us) so that the host will not defend itself against the parasite (the medical-industrial complex). But the host must not be so weak as to die because that would also cause the death of the parasite, and thus a host is usually tricked while the parasite goes undetected so that the relationship can continue. Arnold Milstein, MD, wrote in his January 3, 2013, *New England Journal of Medicine* Perspective article that "Warren Buffet likens our health care system to a tapeworm inside the US economy that drags down our global competitiveness and suffocates funding for K–12 education, basic research, infrastructure maintenance, and other public goods." Of course, Mr. Buffet's interest in such matters is mostly practical, in that healthcare limits corporate profits and therefore the odds of a rising stock market in the future. My interest is different, being a bit more personal, intellectual and physical.

I see an industry that not only drains our nation's resources but also creates more sickness by perpetuating confusion over the fundamental properties of health. For instance, we cannot agree whether we are either generally healthier or generally sicker through our relationship to the medical industry (though the statistics of increasing chronic disease in the population should indicate the truth quite clearly). The US medical system is like no other parasite either: this one not only consumes the host's resources but also transforms the host into another identity — that of a patient — to improve its success.

Modern medicine's power to heal is exceeded by its power to conceal important truths, such as that the best medical management of diabetes (or just about any chronic disease) creates more cancer.

I say this with utter sincerity. It does so by two mechanisms. First, remaining a diabetic, as opposed to preventing it or reversing it, creates immunodeficiencies and allows more cancer to take hold of us. Secondly, the same lifestyle activities that maintain the diabetic state, even while diabetes is being managed well by medications, also cause cancer. Therefore, paradoxically, the treatment of most chronic diseases often produces more disease than it ameliorates, perhaps a different kind. Living with any chronic medical condition, such as diabetes, heart disease, or the other common afflictions, means that we are more likely to face diagnoses of many more diseases than the original one. Subsequently, it could be asserted that multi-organ failure is not only promoted by our culture but by modern medicine as well. Dysfunctional, unhealthy lifestyles create many diseases — and do so additively. Indeed, the odds of a person acquiring a new diagnosis goes up with each new diagnosis despite compliance with the best medical treatments. This is a fact that is not expressed in a typical encounter with our medical providers, wherein they are selling their wares to give us the impression of power.

Medications are premier examples of the reductionist power of the medical industry upon our lives. We have become a pill-for-every-symptom — indeed for every feeling — society, and, by acquiring such a complex industrial identity, we become less resilient people. We become not only dependent upon the Industrial Order but conditioned to this particular mode of thinking. Because this is common in America, as witnessed by the size of our pharmacies and the prevalence of drug ads on TV, it is safe to say that we are trained to respond to life this way. In the old days such behavior was assigned to hypochondriacs, but now that distinction has been very much blurred because industry has trained us to believe in polypharmacy (the habitual ingesting of several medications daily) as a lifestyle.

I often see medication lists of fifteen types of pills per day, meaning the actual pill count consumed is greater. This is a premier example of the dangers inherent to reductionism. The medical indication for each pill on these enormous lists of medications being pumped into people is contrived from extremely sound scientific reductionist reasoning: each ailment, abnormal lab test, diagnosis, and symptom requires a specialized pill to correct it. Eventually, hopefully anyway, our society will be concerned about the mix of medications ingested by a very large segment of our population — but then, when there is so much profit to be had via the status quo, perhaps not.

As our biological identities become influenced by the complexities of these well-defined industrial identities, we are assuming the characteristics of the medications internally, and therefore we are changing as a people. That is, people on medications are more fearful of their health generally and so are less inclined to exercise. For instance, if the diabetic patient exercises, then the medications they are taking for blood sugar control are too potent and they are at risk of diabetic coma. People treated for high blood pressure and heart disease tend to fear strenuous exercise, especially since they were uncomfortable with it before the diagnosis, because they fear a heart attack.

Indeed, most Americans are comforted by their choices of complex foods and medicines. Technophilia (technology loving) and technotaxia (compulsively embracing technology) appear to rule our days and to have overridden our wisdom. By that I mean that our capacity for an industrial identity is best expressed in the consumption of manufactured foods and medicines, the ultimate methods of corrupting a pristine human form. But our physical form is not the only aspect of our lives that is captured and then molded by complex industrial identities, our hope and cognition are as well. Indeed, within our industrial identities is an element of blind faith. Most people are under the impression that medical science operates with precision, but does anyone really know what five and twenty different medications will do to us when taken on a daily basis, no matter our age or for how long?

Our complex industrial identities contain many such assumptions. We assume that medical science and the government will protect our health, but this is only an assumption and obviously not one based on the actual evidence. There is indeed a great deal of life lost through reductionism, although it is generally the life not yet developed fully, the life yet to be lived, the best life that one creates inside oneself. But this is also the life that is unmeasured and therefore unseen, the life that can only be felt. If we are not interested in achieving the best health possible for ourselves, then why would we believe someone else will be interested in fixing us?

Consider the following list of very routine actions that are part of many people's lives, actions by which we can see that the notion of a healthy life has become quite distorted within our modern lives. Being modern consumers, we are not expected to easily see a life dependent upon technology that is simply being sold to us by an industry. It is intriguing to me that most people do not interpret some routine occurrences, these common dependencies and physical consequences that are obvious forms of substantial suffering:

1. More than 150 million Americans are robotically marking the hours of their lives by the pharmaceutical pills they ingest. I frequently ask people why they are on a given medicine, and they do not know. The medication routines they subscribe to are simply part of their personal identities and the exact reasons for taking the pills are less important.

2. Tens of millions of people are checking their blood sugars by routinely sticking their fingers and arms several times per day. By 2050, it is estimated that half of the population will have diabetes, and so this routine is becoming utterly normal in American life. I frequently ask patients what they and their doctors have decided to do about high blood glucose readings, and they say the dose of their prescription would increase or they are to do nothing.

3. Millions of people are drowning in adipose tissue and require CPAP masks and ventilators so that they can breathe when sleeping, even taking these devices with them on vacations. I ask these people if their doctors told them they could perhaps do away with the respiratory support equipment by making some changes in their lives. They tell me that their sleep specialists said that they could recommend a surgeon to cut out part of their throats to let the air flow inside better.

4. Tens of millions of Americans are shuttling between pharmacies, routine medical tests, and doctor appointments every day. Far fewer are doing any biologically purposeful activities such as exercise. Sometimes I wonder if I see more ambulances on the roadways than I see people exercising.

5. One in five American adults (including fully one-fourth of the female population) are on anti-depressants, anxiety medications, and pharmaceutical-grade stimulants to help them tolerate each day in the American dream. Many more people simply choose to self-medicate with alcohol and other recreational drugs. Most of these people probably do not refer to this as suffering either. These patients tell me they suffer less depression if they take their medicine, that is until they need a bigger dose or a new medicine. Rarely do people try to get off their psychiatric medications — and the people taking these medications are mothers and fathers, school teachers, bank-tellers, nurses, and other just regular folk.

The Industrial Order provides us substantial choices, and the extent of artificial interventions in our lives, the degree to which we cultivate an artificial life, is of course determined by the individual, but there is a deeper philosophical consideration beyond the choices we make. Artificial interventions themselves are redefining what it is to be human, and not relative to our own individual merits as organisms but a complete existential translation as dictated by our industries, our machines and our innovations.

Heralding a new human lifestyle is the "Patient-Centered Medical Home," where healthcare becomes an integral part of the patient's home and the patient's life. Like many industrial medical products, it is given a name that imparts a sense of coziness and warmth. In this setting, the patient can remain at home enjoying the endless bounty of industrial-made foods while a team of medical professionals will coordinate everything, such as the timely ingestion of pills, procedural appointments, medication refills, physical therapy, occupational therapy, speech therapy, medical equipment deliveries, device applications, home respirators, and other medical goods. Nurses will be busy administering intravenous drips of medications, while medical professionals in remote offices will be monitoring the patient's blood pressure, pulse, weight, heart rhythm, abdominal girth, and swollen legs. Pulmonologists and respiratory therapists will adjust oxygen levels and inspiratory pressures on the night-time breathing machines. Wound care specialists and nurses will coordinate intravenous antibiotics with dressing changes on the patient's bed sores. If the medical team at the home fails to keep the patient from having symptoms they cannot manage, unstable vital signs or abnormal lab tests, the patient will be transported expeditiously to the physician's office, surgery center, or hospital.

The person will not merely be dependent upon but melded into modern medical inventions. The person's biology will be under direct supervision and control by a professional medical team using the best techniques that evidence-based medicine has to offer. Most intriguingly, the patient will still be enjoying the bounty of artificial foods, maintain the right to smoke cigarettes and marijuana (even on continuous flow oxygen), drink vodka, eat French fries, ice cream, Cheerios, delivered pizza, and other factory-made delicacies. Why should they not continue on the same path that brought them to ill health in the first place since there is the rescue option at hand, and self-destructive behavior is a right and, as such, part of the American way of life?

Complications are a reality for all medical interventions, and so that fact will add to further necessary intervention. Moreover, medical conditions in general are never static but change in aspect over time, and so the above scenario will inevitably include the consumption of ever increasing amounts of medical products. The medical interventions themselves (the procedures causing additional fluctuations in lab data) and the professionals tending to the patient will create additional medical necessities. Meanwhile, the natural aspect of the person's life becomes less relevant as the Industrial Order of food and medicine supersedes any vestige of natural life.

I can imagine a future medical advertising slogan: "Automate your life or die." Actually, this is being said right now, though less directly, but few people are paying attention: "Colonoscopies save lives" and "Call 911 with ANY unexplained chest pain, shortness of breath, dizziness, sweating, or nausea." Headaches can be a sign of stroke, and so of necessity any tingling in the arm might be a stroke these days, and time is of the essence because there is a narrow window to get therapy, and seemingly everywhere I see the words, "Trust your heart to Lipitor." We are constantly reminded of the disease lurking within us, but very little concern is conveyed for the health that is lost.

From such messages a casual observer, let alone an American with a fully formed industrial identity, would conclude that the real hope for our health within modern culture is to place everyone on varying degrees of artificial life support. Medical dependence ensures chronic customers, but that seems beyond most people's ability to understand, as is the implication that patients are required to turn over their health, utterly and completely, to medical professionals and the technology of modern medicine. This plan appears to be compassionate as medical personnel are very busy attending to ailments, but as the process becomes more mechanized, any such claim of necessity lacks sincerity. This plan also appears to be based on sound medical science because the randomized and controlled trials say so, but few ask where the medical industry with its artificial interventions is taking us as a people.

If we become dependent on the medical industry to a much further degree than is already the case, the absolute value of a naturally healthy life may be completely lost to us. This is unfortunate since the direct causes of our most common chronic illnesses are the result of unhealthy habits, which means we can avoid such interventions altogether. Data from the CDC and other reputable groups indicate that the source of diminished vitality and up to eighty percent of our medical conditions is poor lifestyle choices, especially the consumption of manufactured foods and a sedentary life. Yet, in line with the concept of manufactured health, medical institutions espouse a desire to control our symptoms and markers of disease such as cholesterol levels, blood pressure measurements, and other end-points (lab markers for disease). I do not see similar emphasis placed upon the fact that diabetes type 2 is preventable or even reversible, even for those who have a family history of the disease. A person need only shed 10-20% of his or her weight and the diabetes risk is reduced by 60%.

Granted, the baby boomers are getting older and thus more frail, and perhaps they deserve a little technological comforting, but given the trajectory of our current relationship with the healthcare industry (and the proof is that our young people are becoming chronically sick with increasing frequency, developing medical-industrial identities), every American will eventually remain sick through his or her entire life. The previously mentioned advertisements for medical products in fact portend a terrible American fate: future dependence upon the medical industry. In fact, most people I encounter with a heavy medical footprint, such as the routine use of medications, procedures and devices, do not consider themselves chronically ill, further proof of the pervasiveness of the medical-industrial role in our collective industrial identities.

By any objective measure, we have an increasing epidemic of sickness in this country, despite the technological reassurances received in our doctors' offices that our collective future as regards health is rosy — but then few seem interested in objective measures. Sensationalism sells medical products, not realism. We routinely hear fantastic claims of future therapies and cures, most of which are unrealistic in both efficacy and safety. Many physicians and scientists will promote their research to the media far in advance of realistic cures to bring fame, which secures money from venture capitalists or the government. We, as consumers, will never know the difference, and the news they give us will always sound good, especially coming from experts. These curative promises influence our present culture by way of the fantastic hope and false comfort they promise. Always on the horizon of human innovation is something being invented to fix us. We will not need to change our ways because a breakthrough is about to be made, and thus these sensational breakthroughs deepen our apathy and complacency regarding our own health.

It is worth noting that there was a great deal of hope for genetic cures in particular, which were supposedly made possible once we successfully sequenced the entire human genome. However, twelve years have passed without even the slightest hint of potential genetic cures from the genome-sequencing research. I am not saying they will never occur, but millions of people will die of obesity and other more mundane self-inflicted causes before there is one life saved as a result of successful genetic mapping. Genetic diseases are rare, but the man-made environmental effects upon our genetics, such as obesity, heart disease, diabetes, and smoking-related cancers, are extremely common. In fact, I will describe in detail later how our genetic make-up is altered by the manufactured products we consume.

As the overall population is becoming more sick, our medical expenditures are making the entire United States financially sick. The medical-industrial complex is the largest industry in America by far, controlling almost 20% of our entire gross domestic product, more than 2.7 trillion dollars a year — 2.7 times the entire amount of money spent on both the Afghanistan and Iraq wars combined over ten years and four times the entire national defense budget ($633 billion budgeted for 2013 while current defense spending is actually decreasing and medical care is increasing at around 10% annually). Nevertheless, the industry has substantial growth potential because we have yet to see the real burden of chronic diseases. Look around, and you will see enormous factories to make medical products and medical centers under construction, and so the medical industry knows full well there are still resources to capture.

This industrial juggernaut cannot be so much as slowed by the President and the Congress of the United States, the most powerful institutions on the planet. In fact, the government is directly implicated in the creation of the Industrial Order and the conditions that allow its influence to permeate our culture and to thereby control us. Through subsidies of grains and corn, to make manufactured foods more cheaply, to the infusion of vast sums of monies into the schemes of Medicare and Medicaid, the government creates the conditions for chronic diseases to persist and multiply. In fact, this book will argue that individuals will need to lessen the influence that industry has over their health rather than rely on some outside intervention, governmental or otherwise, or be ready for a life of chronic disease. It is only by the actions of individuals refusing to take part in artificial interventions without first implementing natural ones that the overall culture will eventually change.

More often than not, consumerism is prone to weaken us as individuals, but is typically viewed as empowering. The fact that we can choose our poisons makes us believe we are stronger. On any given day, we are barraged by TV ads, radio ads, internet ads, magazine ads, and billboards reminding us that we are a people who suffer from erectile dysfunction, high blood pressure, high cholesterol, low male hormones, urinary incontinence, depression, heart attacks, atrial-fibrillation, colon polyps, and assorted other medical conditions. It would be difficult to imagine that these were all intentional qualities of a natural life, even though some are. The forces we face are, in short, not just powerful but continually assaulting our consciousness. These forces must be faced down or our very human nature is at stake.

The choices we make as adults build the society that our children are reared into. The youngest members of American society are exhibiting inordinate rates of chronic disease and obesity that will cripple them for life. Our children are acquiring an artificial, industrial-made physical existence wherein they are not only obese but suffering from diabetes (by 2008, fully 23% are pre-diabetic or diabetic) and medical conditions such as hypertension and high cholesterol that used to plague people over 65 years of age. These children's clocks ticking toward the end of their lives now start in their elementary school years rather than when they are senior citizens, an unnatural state of affairs, to be certain, and one that can only be created by an artificial existence. These terrible facts serve as proof that it is becoming increasingly more difficult to contemplate a naturally-based, truly organic human life in America, and so we must do it now if it is going to happen.

However difficult, it is not too late. I have found a much healthier lifestyle in middle age, a renewal of sorts because I have considered my health options and found my choices a no-brainer, and if I can do it, so can others. The most devastating quality of an industrial identity is the fact that it conceals our biological identity by diverting our conscious thoughts toward its needs even at the expense of our actual well-being. Within the course of an average American life, we typically lose touch with our natural powers and natural qualities with the early acquisition of an industrial identity, but it need not remain that way. Within the Industrial Order, we can lose the exceptional value of a well-formed biological identity, but that latter possibility is not lost, not yet. I firmly believe that most all of us have fates of better health undiscovered within us. We need only realize the need, then get on the path and commit to it with the utmost sincerity for an exceptional life.

Interestingly, however, when I ask people randomly where the hope is for mankind in the future, the answer will invariably be a peaceful government (as artificial and manmade as any innovation) or a new technological invention, machine, or energy source. People never say that the hope for mankind is in better biology, as in a natural biology as opposed to an industrially created one. This is most interesting because the successful evolution of any organism must entail a better biological skill, adaptation, or physical form. Currently, acquisition of an industrial identity allows one industry (food) to fashion us into manufacturing inputs (unnatural life forms) to serve another industry (the medical-industrial complex). This state of affairs creates physical and mental dysfunction, changes our thinking to suit its needs, alters and therefore controls us. In short, we are looking for health and hope in all the wrong places, and there is plentiful evidence to suggest that the situation is only going to get worse unless we achieve a release from the Industrial Order and the thinking it inculcates. This release will need to be realized at a cellular level.

CELLULAR
CORRUPTION

"The end of the human race will be
that it will eventually die of civilization."

— Ralph Waldo Emerson

Our lives arise from our cells, the most basic unit that harbors the spirit of all life. Even early in high school I was intensely interested in these microscopic entities that somehow account for facets of our lives, a relationship that gets much less respect than it deserves. I embarked on the study of cells in high school, but as my studies became more in-depth in college, I became fascinated with their individual power, resiliency, and purpose. My love of nature was reflected in some unusual ways, and I was not even aware of perhaps having already established such a deep-seated connection to the unseen forces of the natural world. For example, to complete my Bachelors in Science at Florida International University, I was required to take art courses, and so I took Sculpting 101. I did not know what to sculpt, so I tried first to cut a cell out of plaster. It was round, and from a distance there was sufficient definition and attention to organelle detail that the endoplasmic reticulum together with lysosomes and the Golgi apparatus gave the appearance of yin and yang. My art instructor did not fully appreciate my interpretation of a cell in plaster, at least not as much as I did.

Under the microscope, I still find healthy cells aesthetically appealing as well as extraordinarily interesting as units of life. Cells are filled with yet smaller units of specialized function called organelles: the nucleus houses the DNA (deoxyribonucleic acids), and the nucleolus, the darker area within the nucleus, is an area where DNA is being converted to RNA (ribonucleic acids). The RNA then goes out to the ribosomes and they make proteins. The endoplasmic reticulum shuttles the proteins to their intended areas and a Golgi body then modifies them for specialized functions. There are also lysosomes, vacuoles, and a host of other organelles.

Diseased cells, such as cancer cells, will often have very asymmetric, distorted appearances because their organelles are too numerous and bunched up to one side or they are filled with cancer-driven metabolic debris, making them dysfunctional. Cancer cells often look hideously angry under the microscope, and, indeed, cells can actually appear aggressive under the lens of a microscope. The diseased cells will also exhibit distorted architecture by other means, especially relative to function. Cells that make up the lining of one's arteries may visibly contain excess deposits of fat, and they may contain cholesterol if there is atherosclerosis. A fatty liver from being obese and/or drinking excessive alcohol will contain liver cells laden with fat globules. These cells do not look remotely well. Cellular corruption also takes place when the intended function of a cell is hijacked. Because cells are often producing molecules and building blocks for other parts of our bodies, their corruption may lead to excess production or insufficient production of these all-important molecules. In diseases such as diabetes and even Alzheimer's disease, cellular corruption may involve the inefficient processing of molecules. In diabetes, we also see the accumulation of sugars outside of cells, the sugars causing harm; and in Alzheimer's disease we find the abnormal, malformed protein deposits called senile plaques.

Most of us do not think of our lives arising from such small, seemingly insignificant units that, by aggregation, synergism, and communication, provide the means for us to live. Our lives are typically measured in days, weeks, years, and decades, but in fact the biological timeframe, the timeframe at the cellular level, is measured in milliseconds. For example, your eyes are scanning these words with tiny extra-ocular muscles controlling your eye movements and adjusting their contractions in a coordinated fashion by way of neurons. These words elicit neurons to fire first in recognition of the words and subsequently recruit other neurons in a different region of your brain to comprehend the words. As you read these words, biological changes are also taking place inside your brain, changes that are dependent upon and a result of minute quantities of neurotransmitter release and re-uptake between nerve cells. Almost simultaneously, after receiving the letter or symbol information from the retina, the optic nerve transmits the data to centers in the brain specialized for the interpretation of information. Fluently, in milliseconds, opinions begin to formulate as information flows across multiple centers of your brain like a wave. Our lives are truly just a compilation of a multitude of these tiny fluid moments.

Fortunately, it is not necessary to be aware of these biological actions because it would result in information overload. On the other hand, lack of appreciation of these facts can create an impoverished existence realized through physical suffering, physical limitations, and lost opportunities for a better life. People tell me that they wish they could run after their kids and grandkids with more comfort and enthusiasm, and, as a physician, I deal with the epidemic of chronic diseases. All of our abilities are directly related to our cellular health.

It is interesting to think that we are oblivious to the moment-to-moment choices that are frequently harmful to us for the rest of our lives, choices as simple as the food we eat that can corrupt our cells and result in diminished capabilities in our lifetimes. Universally, we tend to believe our everyday actions have no durable impact on our bodies and the rest of our lives, but this is not so. Virtually every action we undertake affects our cells somewhere and somehow. Therefore a relationship with ourselves is really a relationship with our cells.

As the reader has seen, I did not always have the level of respect for these matters that they deserve, despite being a biology major and a physician. But because of some tough physical realities I had to face, I did eventually find a renewed interest in, and a reverence for, the biology of my life. I began to see my biology for what it really is, my hope for life. My capacity to help other people and to enjoy a full life is, after all, a result of cellular functions.

After forty years of age, I had to be humbled into this new consciousness with a shocking realization as I gazed in the mirror and saw a reflection that did not at all resemble the ideal picture of a healthy human. This peddler of health care exhibited an outward physical appearance that most certainly must be the result of severe cellular dysfunction. Based on the macroscopic level, the visual evidence to the naked eye, it was apparent that I had become physiologically and biochemically "corrupted." Obviously, the metabolic and physiologic processes working within me that shaped my middle-aged body image into a form that I thought I was immune from must be important to my overall health. I did not look particularly healthy, and so why would I assume I was healthy? The fact that I had received good news from some biochemical tests as a result of a life insurance assessment a couple of years previously did not make me feel any better about my health. I had no obvious diseases, not yet.

Through the first forty years of my life, I lived as an average American, and as result I became obese, uncomfortable, and, frankly, ill, although I was never diagnosed with any chronic medical condition. Eventually, the discomfort and uneasy feelings I was having as a physician caused me to reevaluate my life and develop a healthy respect for my body, and therefore for my cells. I realized that, in our technology-laden society, it is often forgotten that knowledge of a poor health state does not require sophisticated testing for a cholesterol level, an electrocardiogram, a C-reactive protein level, a physician visit or other medical industry assessment. Instead, individuals in their own private homes can make health assessments quite simply with a bathroom mirror and a scale, as I did. Of course I considered running blood tests and such, but what would they tell me that I did not already know? I did not appear to be a healthy specimen and I did not feel healthy, so why even test for health?

I was feeling older than I knew I should feel for my age, and I felt less vital and restricted in a functional sense. I knew that my life was at risk without invoking the scientific method or seeking a medical opinion. This appeared to be a rather straightforward diagnosis. Intuitively, I knew I could not continue to live as I was and expect good health results or the excellent prognosis for life we all hope for. I also intuited but did not have the words to describe it because I had not completely processed the issues at hand: I had become someone, an unhealthy biological form, that I did not intend to become. Later I realized this newly acquired physical form was indeed another aspect of my industrial identity, my new cellular identity. So, I made an honest appraisal of my health, but what was I to do about the situation? What was the cause of all of this obvious dysfunction? Was age solely to blame, as the people around me said?

My commitment to a healthier lifestyle began with the destruction of some of my assumptions and my routines. I had assumed I would always be healthy, but obviously this was not the case. I had accommodated for this new dysfunctional body while I was hiring yard workers and going to less strenuous vacation settings, and therefore it was easy to feel fully functional in spite of these physical changes. The standard, the bar, for my idea of adequate function and performance needed to be raised. A deep internal change was also required in my mind, hopefully a level of resolve adequate to the drastic change for the worse I saw in the mirror. To say the least, consistent time and energy for self-care in modern American life, including family and career and all else we take on in a given day, can be quite challenging. In fact, there are always a zillion reasons not to engage in self-care but only one reason to do it — a healthy life.

I developed "Charlie's Plan" of diet and exercise, which at first was quite simple: I would eat indiscriminately most of the time and exercise to compensate for the excess calories. I became more active, but at this point in my life, I still did not have a good notion of what constitutes healthy foods, nor was I convinced how unhealthy typical American foods are. I viewed outbursts about the evils of fast foods and processed foods as weird, liberal-minded conspiracies on the part of those who just hate American power and its industrial might. However, Charlie's Plan got a lot more successful as a real plan for health became more refined and I began to utilize the concept of cellular corruption as the cause of our diseases, and even of many of our so-called age-related changes.

Being unsophisticated about the notion of excellent self-care was actually a beneficial place to begin my sojourn, because it allowed me to start from ground zero and build my methods by trial and error. I was medically savvy, even Board certified in internal medicine, but this knowledge framework did not seem to improve my odds of establishing a higher performance standard and a better standard of health for myself. A person's education level does not override the hurdles and complexities of modern life with regard to maintaining healthy habits. I realized that I needed a desire for a healthier life and the grit to carry it through, that in fact intellect was really not needed, just action and more action. I realized too that there is a certain amount of enjoyment in the accomplishments one achieves through trial and error, but more importantly, that changes are more likely to occur by the methods that one develops for one's self, that health should be simple and easy, and so the plan to pursue it is best found by finding what works for you. This way the person begins with an emotional investment of sorts. In few words, curiosity can motivate and success derived from one's trials is the best reinforcement.

As I began to change my diet, I had less of an opinion about what I ate than of the quantity I ate because I was still under the impression that all foods were equally nutritious. I still believed that all food calories were equal in their effects, that "a calorie is a calorie." Being naïve about the real concerns of health, such as a proper diet and genuine self-care, positive changes to my overall health came slowly as convenience and cravings for the usual American fare continued to prod me. Industrial foods such as Doritos, Fritos, Hot Pockets, microwave burritos, and reconstructed sandwich meats like hot dogs and bologna seemed necessary to everyone's busy life, and I was the epitome of this thinking. Also, at the time it was fun to eat these foods with my young children, and as they grew they expected them as well. These early days were marked by restrictive eating as measured in calories and portions, hence extremely small portions for the high caloric density of modern food. Then there would be spells of renewed gluttony followed by punishing exercise routines as compensation.

I could not wrap my head around what I really needed to do to regain better cellular function and that elusive goal of optimal health. With many details missing from my experiments, further resolve was necessary to begin to meticulously log my food ingestion and my exercise habits, and so I started planning meals and activities. Meal planning can provide confidence and control, I realized. For instance, it made me feel good to have one "lose it day" per week. Being somewhat knowledgeable of physiology and experimenting by my own means, I even rationalized "lose it days" as necessary to diminish the expected metabolic complacency my body's machinery would be expected to feel if I were a sane eater constantly, a predictable creature. Our bodies are very adept at maintaining our weight; therefore I rationalized I must vary the nutrient supply to prevent weight-loss stagnation. This seemed logical to do.

I would eat anything I wanted on those days, but I would still keep track of my caloric intake for the scientific and curiosity value. I recorded days of 3500 plus calories, which were obviously not healthy or necessary, even under my initial rationalization. Still, I was reassured in that I was acquiring knowledge about myself and by the fact that I remained committed and headed in a new direction. At a minimum, I was becoming aware and not simply passing my time with the truth utterly obscured by my industrial identity. However, these behaviors caused me to yo-yo in weight.

I am a scientist, and so I plotted my weight on graph paper both when I was following my diet plan and when I was not. Sometimes it was fun looking at my weight and what I ate, say, when I added up a day's worth of calories and was pleasantly surprised at the low number; but more often, I was astonished and discouraged by high totals. In other words, there were trends, but it was not always easy to predict on a daily basis what my interventions would do to my weight. I will say that it was always useful to see bad results in this format, however, because I was strongly prompted to return to sanity by objective data. I achieved some initial momentum based on my positive results both on the scale and in the mirror, and I became convinced I was on the right path. These results in turn led me to become more motivated, interested, and invested in my process. But my results did not give me the feeling of leading a life in balance. In fact, I always felt I was at war with something, and I know now that the war I raged was with my old notions, born of the industrial identity I had acquired over decades. I am stubborn, as many Americans are, and I found it difficult to give up concepts such as eat more, do more, accumulate more — it's the American way. Just do it! However, as noted in the previous chapter, it is not as easy as willing one's self to a different view of health by virtue of our American environment and culture of excess.

I created the 1200-calorie day, then a 1500-calorie day, followed by an 1800-calorie day and a return to the 1200-calorie day. The weekends would contain one or two days when what I ate did not matter (debauchery days). I thought there could be value in varying the amounts of food so that I would have more flexibility with my meals, and I reasoned that I was less likely to be bored with my eating patterns. In my plan at this stage of its development, it was okay to have days of indiscriminate eating. I would make up for my momentum toward health with a punishing workout and by eating less. This system did not necessarily lend itself to consistent nutrient supplies, however, and by being sometimes so calorie-restricted through the day, I could not help myself from eating indiscriminately later at night. Because I was keeping records, I could see that I was compartmentalizing bad behavior. The holidays proved prime examples of such behavior. The most important lesson I got from my initial experiments was, of course, the importance of real honesty with oneself, living in a world of endless food opportunities.

A few concepts I learned during these trials are worth mentioning:

1. Honesty can be shocking but is most definitely necessary for one's health. Without honesty one cannot effectively change the aspects of life that are detrimental.

2. I had no idea what I was actually eating before my record keeping. On paper it became quite obvious that I was living to eat rather than eating to live. My exercise habits were adjusted often because of my eating patterns rather than being performed to be simply healthier.

3. Although record keeping may be tedious, it is a very beneficial tool for learning about oneself and for discovering what works. Because I kept track of my food intake and exercise habits, I could place value on various interventions. For instance, it became really obvious that what I put into my mouth mattered to my physique and my overall physical condition. It is important to know how much eating a whole pizza or bowl of brownies affects one's body. Indeed, I was not aware of the magnitude of cellular corruption I was capable of until I kept records and saw the numbers in black and white. No microscopes or exotic blood analyses were needed.

4. Meal planning has virtue in that I am able to maintain what is important to me, to control for environmental conditions, namely living in a society built on excess food, and to have confidence in my life. Planning allows me to eat what I prefer rather than eating whatever is

available. Consequently, I plan for many aspects of my day and none possesses more importance than my nutrition.

Over time, I managed to escape the food habits and food philosophy I had acquired in childhood and had lived by since, but first, with improved discipline in my planning and record keeping, I slowly achieved decent weight loss. But I was not yet focused on the value of these interventions for health. Vanity is a powerful motivator, but as I began to feel better and perform better physically, I became more invested in the purely health aspects. I certainly did not wish to end up as someone's patient soon, and this is the point at which I began to analyze the habits I had in common with people at risk for chronic diseases.

My diet, devoid of fresh fruits and vegetables and consisting of energy-dense foods with excess calories, was setting me up for the "industrial diseases," conditions like Alzheimer's, cancer, vascular disease (stroke, heart attack, peripheral artery disease), and diabetes. I could recall very clearly what an angry cancer cell looks like under the microscope compared to a healthy one, and it is well-known that a diet substantially of vegetables lessens one's cancer risk. As I had seen in my own data, the choices of foods I put into my mouth altered my body anatomically, but I also knew that detrimental health effects are occurring within people long before the person is diagnosed with these diseases, and my own data reinforced this notion.

But one's food choices alter one's psychology as well. Unfortunately, an altered mental reality (via effects on brain chemistry) translates into an altered physical reality (changes in one's health and body size) in a kind of closed-loop feedback arrangement. I have often jokingly said after eating large amounts of manufactured foods that I felt like a beached whale, and this was an extremely accurate feeling, based on scientific understanding. My physical identity was being molded by an artificially-contrived food psychology.

Clearly, our physical discomforts are overridden by consuming more food from a factory and placed into a box where we find incredible convenience. We can sustain comfort from a myriad of sources, but food is certainly one of the most consistently comforting items and the most available. Many people are in fact lulled into apathy because these foods act like drugs in our brains, these manufactured foods stealing our health by making us feel comfortable.

Curiously, foods from factories can make us feel comfortable even when we waddle or our pants feel tight as our bodies change into different shapes.

The effects of manufactured foods on our minds and bodies are surprisingly powerful. The industrial identity I am encouraged to have with regard to non-edible goods pales in effect to a changed physical form. Manufactured foods create an industrial identity at the cellular level which confuses or corrupts our essence. For many years I intuited that the foods I was eating were not the best choices for me, but instead of researching the basis of that intuition, as I do with most important things, I continued eating these foods under the influence of the delusion that they would provide the same nourishment as natural foods. After all, they were fabricated from particles found in nature. However, when I finally did awaken from my industrial slumber sufficiently to investigate their potential nutritive value, I discovered that, although the atoms may have come from natural forces, processing them into a breakfast cereal, muffin, or granola bar alters the way these molecules are handled by our bodies. The presentation of our foods to our bodies is extremely important to our nourishment.

Nutrients must be presented to our cells and then absorbed by precise mechanisms that have evolved over the eons before modern society existed. After absorption the nutrients are transported and processed by various organs, such as the stomach, intestine, and liver, before incorporation into our tissues. The proficiency by which we digest, absorb, and assimilate nutrients is matched precisely by the architecture and constituents of the foods being produced by nature. This cannot be said with regard to the foods that are made in factories. We risk a great deal by believing what is printed on the boxes and wrappers of manufactured foods as regards their goodness. Marketing claims aside, a significant loss of nutritive capacity is established as the molecules are stripped of their vitamins, minerals, and other micronutrients during processing. To make matters worse, the reassembly of food molecules into consistently perfect forms of tarts, pastries, bars, chips, crackers, breads, TV dinners, microwavable meals, juices, syrups, and other artificial items alters the food molecules in such a way that it induces adverse reactions in our bodies, giving rise to the epidemic of chronic diseases. As I will elucidate further in later sections, manufactured foods pose an extraordinary threat to us, one at the cellular level.

The obesity epidemic can only be sustained by one aspect of American culture: the industrial-made diet. After all, if I as a physician with a lifelong passion for the study of biology could be so blinded to the effects of manufactured foods on my health, then this must be true exponentially for the majority of Americans. As I slowly became more in tune with my biological existence, I began to sense barriers to my health that were not initially identifiable. I realized that, therefore, a thorough review of the basic tenets of my life was in order if I was to achieve excellent health, the best life possible, rather than the one society was pushing me to have. During the times I felt more desperation to change my course for health, my original affinity and admiration for the life of cells was reinvigorated. I began to recall those passionate moments during my education years in which I marveled at the complex potential held in every moment of a natural life, the life of cells.

MOMENTS MATTER

Our cells live in the moment, and this is worth remembering. Our cells might be primed for future events through adaptation, but they do not predict the future. It could be said cells remember past events by virtue of the mark of some previous event left upon them, but they do not dwell on them. Our cellular moments determine the overall quality of our lives, our health destiny as it were. Likewise, the contents of our minds are also dependent upon the cellular health of our neurons and our actions. Even our joys and sorrows are cellular events, and thus, the importance of our cellular health should not be under-appreciated.

We can choose to bathe our cells with nutrients aligned with our biochemical heritage, biochemistry/physiology created by natural means, or we can choose not to. Our compulsions to eat, our tastes, our post-prandial feelings, and our preparedness for current and future activities are choices that represent our priorities on the macroscopic scale, but ultimately our choices affect our cells, which in turn truly affects us at the macroscopic level by determining the health of the organism as a whole. In order to understand health and disease, we must appreciate the intricacies of our cellular moments because our food and activity choices, although macroscopic, define the conditions for the microscopic origins of our lives. We as individual organisms reside at the point we do on the life continuum between health and disease by virtue of the moments our cells "experience," those moments adding up to our overall physical and mental health or the lack thereof.

I developed an appreciation for the intricate nature of our biochemical fates while I was preparing for a presentation on the human clotting cascade during my Internal Medicine residency, an important moment that nevertheless slipped my memory for several years while I made my way toward cellular dysfunction. Blood clots are formed by what is called a cascade reaction. An initial insult, a cut or disruption of the blood vessel, causes platelet plugs to be activated and to link in response to a break in vessel integrity. During these moments, platelets morphologically change and cover the hole in the blood vessel. Further activation proceeds with a cascade of several clotting factors and tissue factors that reinforce the platelet plugs, effectively sealing the gaps. There are twelve known clotting factors, mostly enzymes, which activate one on another in a specific order so that the end result is a stable plug for the hole in the blood vessel. A mesh of connective tissue is then laid down so that the platelets can remain long enough for the vessel to repair itself. These actions take place in milliseconds. If insufficient clotting occurs, we run the risk of hemorrhage and potentially death. If too much clot is formed, then we run the risk of other adverse events such as a pulmonary embolism (a blood clot to the lungs), stroke (80% are due to blood clots in the brain), or heart attack (a blood clot in the arteries feeding the heart muscle). Clotting mechanisms must therefore be precisely controlled every second or we suffer.

The clotting system is constantly poised to clot and/or dissolve a clot at the very same instant, a condition that is fluid, constantly changing in response to local and systemic changes. That is, in the same milliseconds that clots are forming, enzymes are dissolving the clot. For instance, at the periphery of any clot there are tissue-specific enzymatic and humoral influences that sculpt the size and shape of the clot being formed. This clot remodeling occurs in every artery, vein, and capillary throughout our bodies. These incredibly intricate and complex processes entail simultaneous adjustments and occur continuously for the duration of our lives, an elegant and essential cellular dance within human physiology that might be termed a set of perfectly attuned cellular moments, when the system works as it should.

These physiologic matters are made infinitely more complex by the fact that blood is a constantly moving tissue. Regionally specified metabolic demands are placed on it, and our blood responds with instantaneous biochemical adjustments as it continuously moves throughout our bodies. These demands are extremely variable, depending on an organ's activities. In the brain, a blood clot is extremely serious (stroke), and the capillaries that nourish our brain cells are very small, so it is advantageous for our blood to be on the thinner side and less prone to clotting. Therefore, in the most fantastic fashion, our blood becomes a little less prone to clotting immediately upon entering the brain's privileged sites, which are behind the curtain known as the blood-brain barrier. This is in some ways a physical wall but in other ways almost a philosophical wall, a special milieu created by a curtain (metaphorically) constructed within the walls of the blood vessels that nourish our brains. Here biochemical events happen that are not replicated elsewhere in the body. This barrier filters and adjusts blood constituents, metabolizes drugs, guides immune responses, and also sets the clotting cascade mechanisms to a thinner predisposition.

Upon exiting the special brain space, our blood is physiologically/biochemically adjusted in drastic ways, precisely reversing the changes that occurred on the way into the brain space, including again becoming more prone to clotting, as that is what is important to other regions of our body. This incredible biologic milieu surrounding our central nervous system continues from our inception until death. Blood is a flowing form of extremely active tissue we can easily forget about as we move through our days, and yet the physical properties of this tissue are not static in nature. Therefore, the power and potential to give us life is derived from a finely coordinated dance of opposing molecules and molecular forces. This dance of life provides for our excellent resiliency overall.

A lack of appreciation for these beautiful events breeds threatened moments for our cells, apathy toward a natural life, and overall risk for the overall organism. In an instant we can experience a heart attack or stroke, but the reasons are far from immediate. The process that leads to these negative outcomes, these critical care emergencies, is the accumulation of destructive biochemical moments over years or decades, biochemical events that are largely dependent upon how we live our lives. The Western diet, smoking, and a sedentary lifestyle cause atherosclerotic plaque to build up in our arteries. This causes narrowing of the diameter of the arteries, and as a result the flow of blood is diminished. With slower blood flow, we are more prone to angina (chest pain associated with heart disease) because the heart tissue cries out when it lacks the blood it needs to carry on its function, to pump in order to meet the body's demands. As plaque builds up, the arterial walls are less compliant and less responsive to augment flow for any increased activities. They are narrowed, less functional, and more prone to being clogged.

The Western diet also leads to cholesterol deposition within the macrophages (clean-up and immunity cells that line the arteries as well), interestingly called foam cells because they look like foam when they are laden with fat, their appearance altered just like the appearance of the whole body. Macrophages are immune cells that eat and digest unwanted debris and virulent invaders that have been neutralized by other immune system cells. When they are active and cleaning up things like cholesterol plaque that is not supposed to be present in the arteries, they will typically elicit an immunological response and therefore inflammation is associated with them. Due to these digestive functions, the cholesterol-laden plaque becomes softer and more prone to break off in parts. This is why it is not so much the calcium-laden plaque, the harder plaque, that leads to heart attacks.

Plaque rupture causes most heart attacks because the sudden event, torn tissue, sets off the clotting cascade, and subsequently blood flow diminishes as the artery is suddenly shut down. As the heart muscle starts to die during the heart attack, pain in the chest can become intense and the patient typically becomes sweaty, pale, nauseated, and short of breath. This dreaded event is the culmination of years of cellular moments. Foam cells do not occur overnight, so, although the heart attack is sudden, this process takes years (even decades) of repeated unfortunate biochemical events, moments of macro-level dietary whim that result in moments of cellular corruption that all stack up over those many years to result in a organismal cataclysm. We used to believe that much of the adverse chemical changes taking place as result of the Western diets were a result of years of ingestion, but we are now seeing immediate changes in our biochemistry (changes within the cellular moment, as it were) as a result of this diet that, over years of ingestion, ultimately do induce major medical conditions. We can obviously sustain a great deal of bodily injury without even being aware of it.

Within minutes of eating a meal consisting of processed, manufactured sugars (even bread, pasta, and cereals), for example, there are measurable effects within our bodies. Inflammation cascades associated with insulin spikes are well-described in research literature. Likewise, within hours of eating a meal high in saturated fat, especially trans-fat, researchers can detect pro-inflammatory markers of cellular communication along with sticky platelets that promote clots. Blood vessels lose responsiveness by way of disordered metabolism of nitric oxide within the blood vessel walls. Again, they lose capacity for regulating blood flow appropriately, and these serious changes are compounded by others, such as an increased level of bad cholesterol (LDL and triglycerides) and lower levels of the good cholesterol (HDL), which is produced by the liver and protects us from atherosclerosis by acting as an inflammation scavenger. HDL molecules neutralize the potential harm of oxidized (angry) LDL molecules that cause more inflammation. Therefore, a lower level of HDL cholesterol is quite detrimental to our vascular health. Excess inflammation for any reason causes us to age more quickly and become chronically ill. Our tissues become corroded and dysfunctional as we are awash in sticky, leaky, angry cells. All of this by virtue of eating saturated fat, which is in many of the manufactured food products we mistake for actual sustenance.

BURNING-UP INSIDE

Inflammation gets a bad rap these days because Americans tend to engage in lifestyles that continuously corrupt this extremely important systemic response, in effect turning it against themselves. Inflammation is not bad in and of itself, but excess or inappropriate (chronic) inflammation can be very detrimental, even lethal. Functioning normally, the inflammatory pathways are intended to fight cancers and the invasion of foreign organisms (viruses, bacteria, fungi), initiate the clotting cascade, and heal our wounds. Like the clotting cascade, there are a myriad of cellular intricacies and intimate biological relations within the inflammatory system. Complex coordination and communication is arranged between several organs, including the lymph nodes, bone marrow, blood components, vascular lining, nervous system, and spleen. These connections occur over various regions throughout the body, and coordination of the inflammatory and immunity responses does not simply occur at the site of injury. Threats and injuries are sensed throughout the body, using various cytokines and complementary cell components. These are relatively small molecules from the classes of interferons, interleukins, prostaglandins, Cellular Adhesion Molecules (CAM) and the Tumor Necrosis Factor (TNF). These proteins and glycoproteins interact with cells such as the Natural Killer cells, macrophages, T-cells, B-cells and neutrophils by binding to specialized sites that translate into powerful biological effects. The inflammatory/immunity pathways both stimulate and coordinate an extremely powerful response to combat microbial intruders and cancers.

Recruitment and delivery of cells escalate to wage a war against intruders, but all of this fighting leads to tissue destruction that threatens survival. Therefore, the clotting and healing pathways are also simultaneously activated for injury containment, waste removal (dead soldiers and intruder debris), and repair of the normal tissue architecture. Regional blood flow is enhanced by prostaglandins and leukotrienes so that the cells involved in the fighting and repairs can arrive promptly at the site of injury. Meanwhile, more precise measures to control the vascular bed is provided by nitric oxide, so potent that it is only released in minuscule amounts by endothelial cells lining the blood vessels. The nervous, immunologic, endocrine, and cardiovascular systems are some of the organ systems that synchronously control blood flow to restrain these intricate processes so that they remain at the site of conflict or injury. Inflammation, immunity, and repair require a wide range of cells coordinated by different organs and regions of the body to fight, clean-up, and repair our tissues within the span of milliseconds. These events are likely occurring somewhere in the reader's body at this very second.

Excess inflammation does not serve a useful biological function and can be created by the foods we eat (direct stimulation) and a sedentary lifestyle (disordered energy balance). If there is no injury or target, such as a bacteria or cancer, then our inflammatory pathways will turn on us and destroy us from within. In the interest of succinctness, I will describe several independent mechanisms by which manufactured foods cause excess inflammation, but not all of them. There are not only more, but the situation is extremely complex because each of these independent mechanisms also worsens the effects of the others; the interactions of manufactured foods with our bodies are compounded and thus doing exponential harm.

First, unhealthy manufactured foods come to us in many forms and tend to exhibit a predilection to abruptly raise our blood sugars, which is quite detrimental to overall health. We are not bred to accommodate these products, which from an evolutionary point of view are new toxins. Examples of foods that cause excess sugar levels are breads, chips, cereals, bagels, pancakes, waffles, biscuits, cakes, donuts, tater tots, hash browns, French fries, and candies. I will explain in more detail later, but none of these foods is older than roughly 5,000 years, which means they are not accommodated well within our genetic make-up. More recently, the invention of high-fructose corn syrup (HFCS), which is found in many foods such as soups, TV dinners, juice drinks, soda products, and even condiments, notoriously corrupts the elegant inflammatory pathways and has been shown to promote inflammation throughout the body by spiking our blood sugars and thereby activating cytokines such as interleukin-6 and TNF. These cellular messengers are also implicated in the processes that lead to aging in general because they tend to circulate at higher levels as we age. Hyperglycemia impairs our cellular immunity because the cells we rely on, neutrophils, lose their mobility (chemotaxis) and power (degranulation of fighting enzymes impaired) to neutralize the invading organisms. Manufactured foods also tend to dump saturated fats and free fatty acids into our bloodstream, and these are pro-inflammatory molecules as well. Overall, our cells get sticky with excess sugars attached to them, just like molasses.

Second, excess blood sugar induces excess insulin because this hormone allows the sugars to enter our cells. We are inclined to deposit fat with high sugars because insulin promotes fat deposition, but insulin also contributes to cancers because of its anabolic and pro-inflammatory effects. Third, manufactured foods are generally calorie dense and therefore contribute to obesity, a dysfunctional state in and of itself. Obesity compounds the overall inflammatory situation independently because it is associated with higher circulating insulin levels through the mechanism of insulin resistance. Our bodies sense this state of affairs and compensate by producing more insulin in an effort to force sugar levels down. Overweight people tend to maintain excessive insulin levels, and this hormone becomes less effective because the receptors to which it attaches become less responsive. Being overweight also means there are excess circulating free fatty acids and abnormal cholesterol patterns also promoting tissue inflammation. Elevated insulin levels combined with elevated sugars and free fatty acids are the hallmarks of both the metabolic syndrome and diabetes type 2. Further compounding a very unhealthy situation is the fact that higher insulin promotes more fat deposition and the cycle viciously worsens.

Manufactured foods corrupt our brain chemistry as well. The cycle outlined above feeds (no pun intended) itself because our brains crave manufactured foods. Calorie-dense foods saturated with processed grains, sugar, and fat stimulate receptors that promote a feeling of comfort that can become addictive, inducing "cravings" much like other drugs. As it turns out, the endocannabinoid system is quite complex and responsible for control of our energy homeostasis and metabolism as well. Activation of the endocannabinoid system (ECS) promotes fat deposition (even in the liver), cholesterol production, and insulin resistance. It also decreases muscle metabolism, energy expenditure, and satiety, thereby promoting higher insulin and sugar levels.

The hormone leptin has recently received a great deal of attention because it was found to increase in response to a meal and give the person the feeling of satiety. Unfortunately, this is only accurate in thin people because obesity is associated with leptin resistance similar to insulin resistance, wherein leptin is not able to shut down the appetite effectively. Leptin is a hormone produced by adipose tissue, and therefore, the more adipose we have, the more leptin that is circulating in us — but it is corrupted leptin and not able to serve its intended purpose. Thus, manufactured foods induce excessive eating through the ECS while at the same time causing a dysfunctional braking system in leptin. I wish I could stop here with the bad news about the negative effects of manufactured food on the human body, but there is more.

Advanced glycated end-products (AGE) are a relatively new health threat associated with most of our nation's epidemic diseases, such as cancer, diabetes, and heart disease. AGEs are molecular by-products of food manufacturing processes that cause us harm by inducing intracellular switching mechanisms that contribute to the production of inappropriate inflammation and diabetes mellitus type 2. Their effects are significant in that they are implicated in altering the cellular signaling that governs the condition of our vascular system, all the way to the genome. They are also potent stimulators to increase Insulin-like Growth Factor signaling, an important mechanism that induces cancer growth. AGE accumulation has also been shown to induce senile brain plaques (tangled, malfunctioning neurons and proteins), causing Alzheimer's dementia. Although fried products such as French fries, chips, and virtually all pre-packaged foods are by far the worst offenders, AGEs are formed when carbohydrates such as wheat, oats, or potatoes are heated to temperatures that cause them to brown or become crisp. A table of AGE-laden foods can be found in the Appendix.

There is a spectrum of disease and dysfunction with regard to manufactured foods, but increased exposures are associated with the increased likelihood of disease. For instance, the insulin-carbohydrate response in the general population is a continuum, and therefore progressive. Slowly, not overnight, people become overweight, then obese, then develop insulin resistance, then the metabolic syndrome, and, if the situation is not corrected, eventually they are at risk for diabetes type 2. One can see that insulin resistance is really a rehearsed state of biochemical affairs, meaning it can be halted if we stop practicing a lifestyle that promotes it. When people are lean and active, blood sugar control is very tight because the insulin receptors on the outside of our cells are sensitive. While I was in post-graduate training, type 2 diabetes was regarded as adult-onset diabetes. Unfortunately, now we diagnose type 2 diabetes in children and young adults because we are undergoing significant cellular corruption from the time we are born. We choose the foods to eat and therefore train our normal hormone systems to be dysfunctional. The earlier the training begins, the earlier we see the disease taking hold. I am confident we have yet to fully understand all of the negative ramifications of the American diet.

I explain to my patients that excess inflammation is like having background noise or constant static in our systems. With the heightened level of alert associated with chronic inflammation always in place, immune and repair resources are diverted from worthy causes (intruders, healing) and will not be available when needed. Also, with the constant background noise disordering cellular communication due to the excessive activity of the cytokines, our ability to detect a real threat becomes all the more difficult. Our systems cannot detect a cancer or infectious intruder because the cellular signals associated with the threat will be lost in the cacophony of inflammation.

Obviously, industrial-made meals are perceived as common threats to the human body because they invoke excess inflammation/immunity, as if these foods were sinister invaders or toxins, but our daily ingestion of them means we are being attacked chronically. Interestingly, removing these products or toxins reduces the immune reaction (even foam cells in the coronary arteries, etc.) and the number of angry molecules coursing through our bloodstream. Things calm down, and our diseases abate.

The distorted, less natural body image apparent when we are overweight is a good indicator that our symphony of cells has become discordant, because the untoward biochemistry occurring within us at a cellular level eventually shows on the outside, but as this discussion of cellular dysfunction in the presence of manufactured foods indicates, the damage on the inside is going on from the day we take our first bite of industrially produced cuisine. The fact that something as important as an inflammation/immunity system is jacked up without a good purpose indicates not only that the system has lost its legitimate purpose but that our diet has also lost its purpose by becoming anything but nutrition for our cells. These cellular level changes do not just lead to their externalization either. Certain diseases will capitalize on a body that has become corrupted. Cancer is one of them.

CANCEROUS CONDITIONS

Cancer epitomizes the concepts I have laid out with regard to cellular corruption. It is perhaps helpful to view our bodies as the soil that can repel cancer or promote the growth of cancer, a description first put forward in 1889 by Stephen Paget who was interested in the biomechanics of metastatic cancers. First, there needs to be a detrimental alteration in the DNA of the cell that transforms to a cancer cell. The alterations in DNA structure can arise from spontaneous or random mutations, or they may arise from acquired sources. Some of the acquired states that are cancer-prone are internal derangements, such as being overweight, chronic generalized excess inflammation, diabetes, and autoimmune conditions. Other acquired conditions may arise from external sources that include tobacco product poisons, ethanol (alcohol), and ionizing radiation (UV light from sun and x-rays). Second, there must also be dysfunction within the DNA repair mechanisms so that, once the cancerous DNA changes occur, they are allowed to persist because they are not excised. Without deletion of the damaged DNA, the cancer cells tend to fulfill their intention to live independently from the body. They actually go "off the reservation," becoming unruly and operationally defiant to the messages from the surrounding tissues to remain controlled in their growth. Cancer cells are corrupted architecturally and as regards function, corrupted from their original purpose as cells. Cancer will only take hold in the body if the conditions are ripe for it and the immune defense mechanisms are remiss and not able to identify it as a problem. There are many steps for a cancer to take hold, and thus many opportunities for bringing to bear more favorable cellular conditions for ourselves.

Malignancies are by definition cells that have become so corrupted from their original purpose that they invade and injure the surrounding tissue. Cancer cells serve only themselves. They have no larger purpose. Losing their way by virtue of the alterations in their DNA template essentially converts them to an extremely selfish form of cell that does not concern itself with the activities of the rest of the body. Instead, the cancer will coerce normal cells to serve it. Malignancies will hijack vascular resources, nutrients, and our systems of resiliency to further its growth. In short, a cancer cell tends to steal life, but only if it is aggressive enough to steal it from the existing tissues. Inflammation plays a definitive role in the development and metastasis of cancer. Some have assigned the inflammatory response a name relative to this disease: flame of cancer. As I have described earlier, inflammation promotes cellular dysfunction through disordered cellular signaling, vascular leakage, and ineffective defenses against these villains from within us.

Not all cancers are malignant, and therefore not all cancers will invade other parts of the body and steal one's life. Cancers are heterogeneous in cell structure and biochemical abilities, and so to some degree they will generally reflect attributes of the tissues from which they arise. Some cancers, such as prostate and basal cell cancers of the skin, are extremely slow-growing and not necessarily invasive and therefore may or may not impact a person's life at all. That is why there is so much controversy over whether or not to screen every male over 50 for prostate cancer, because most people can have it but die of other common causes of death, such as heart disease and strokes, years before the cancer ever hurts them. The majority of cancers that are created are never known because the soil repels them. Our epidemic of cancer these days may not be the sole result of exposure to cancer-causing toxins as much as of the fact that our tissues are so corrupt that they do not repel them.

It is important to understand cancer fully for health reasons, because it will become apparent that we as individuals have substantial power over our fate even when it comes to cancer. Understanding the mechanisms of cancer also serves a larger purpose for discussion because it presents an opportunity to introduce important concepts for the rest of the book. Through an understanding of cancer-related issues, we can better understand how our lives are an absolute function of our cells. Our destiny is a product of the soil we create.

The primary defense against cancer is avoidance of particular states (excess weight, ignorance — higher education level lessens cancer risk), toxins (cigarettes, excess alcohol, processed sugars, fat, pesticides), infectious diseases (human papilloma virus, human immunodeficiency virus), and radiation, which promote cancerous changes in our cells. Smoking tobacco exposes one to cancer-causing toxins, such as benzo-alpha-pyrene, that directly alter the DNA in our tissues. Smoking raises the risk of most cancers and not just lung cancer, but virtually all cancers. The resulting DNA changes caused by tobacco products induce genomic instability, including dysfunction in our DNA repair capabilities. Therefore, smokers are more prone to disordered cell transformations (cancer), but a smoker's exposure to cancer risk does not end there. Smoking tends to change the architecture of our lungs, similar to the effects of obesity in that our normal cellular architecture (the soil) is corrupted. With this dysfunctional architecture, immunity and healing cannot occur as intended and therefore we are simply more at risk. The detrimental effects of the toxins in the DNA are compounded by disordered repair and immunity.

However, most people are not aware of specific health conditions, such as being overweight, excess hormones, chronic stress, and lack of routine physical activity, that are also known to either cause cancer or at least be associated with a much higher risk of cancerous transformations. Being overweight causes genomic instability by causing total body inflammation as pointed out above, but it also changes the body, making it less capable of mounting a defense against cancer. The National Cancer Institute reports that overweight Americans constitute 14% of men's and 20% of women's cancer mortality. Adipose tissue injects hormones into our blood stream (estrogen, leptin and others), and these hormones destabilize many aspects of our bodies, not just our center of gravity and how we view ourselves. According to experts at the American College of Gastroenterology, the number one risk factor for women to develop colon cancer is obesity (ACOG meetings 2007). Excess body weight is clearly associated with increased risk for the following cancers: breast, colon, endometrial, gallbladder, adenocarcinoma of the esophagus, and renal cell carcinoma (kidney). Excess body weight is shown to have a high likelihood of contribution to the following cancers: pancreatic, gallbladder, thyroid, ovarian, cervical, multiple myeloma, Hodgkin lymphoma, and aggressive prostate cancer.

As we assume an industrial identity, a cellular identity that is different from our intended biological identity, the healthier version of ourselves, then we are at risk of disease. Manufactured foods lack authentic nutrition and therefore make us prone to acquiring the physical attributes of the industrial age, a new form of suffering. Corruption of our cells is corruption of our lives in that we miss opportunities for a more vital life.

Missed Opportunities

It is human nature to evaluate cause and effect when facing a problem, but what is as important in many instances, certainly with regard to our health, is what we are *not* doing that contributes to the rampant spread of certain diseases through the American population. Our lack of exercise, for example, an important aspect of the Industrial Order within which we live, is one of the sources of our diseases, and given that physical movement is part of our genetic heritage, to *not* perform essential functions in life means we are unraveling our own biology.

In fact, research has shown that routine physical activity decreases the risk for the following cancers: breast, colon, endometrial, and prostate. According to the National Cancer Institute, routine exercise is known to decrease one's risk of colon cancer by 50%; and it is also known that a sedentary existence elevates the risk of almost all cancers. The *British Journal of Cancer* compiled 20 previous studies of the effects of a sedentary lifestyle in 2011. They found that precancerous conditions such as colon polyps were 40% more likely in the people lacking routine exercise.

Physical activity contributes to a healthy energy balance and is therefore crucial to life, and so the need for physical activity seems self-explanatory. One important aspect of human evolution is our physical being and all that the environment in which our ancestors inhabited inculcated into our DNA in that regard, and therefore it should not be a surprise to us that a lifestyle that shuns the virtues of physical activity will eventually corrupt our cells and allow cancers, perhaps the ultimate punishment for a lack of a natural biological identity to take advantage of the circumstances. The fact that cancers are associated with a sedentary lifestyle can be better understood when we acknowledge the human body as a resilient biologic fabric that must contain energy related to purpose.

I will speak to the absolute need to govern our energy well in life in later chapters, but for the moment, this assertion regarding physical activity will suffice: not doing exercise routinely is very detrimental to maintaining healthy tissue architecture. The Harvard School of Public Health estimates that up to 75% of American cancer deaths can be prevented, and one's energy balance, body size, and activity level obviously play an enormous role. "Exercise should be a 'standard part of cancer care.' Exercise can reduce the risk of dying from cancer and minimize the side effects of treatment. Evidence shows that physical exercise does not increase fatigue during treatment, and can in fact boost energy after treatment. It can reduce the chance of dying from the disease and may help to reduce the risk of cancer coming back by 40% (breast) and 30% (prostate)" (from BBC News, 7 August 2011).

When we possess healthy bodies, tissues, and cells, cancer begins its journey in a very hostile environment, a backdrop of vital fabric that is likely to discourage its growth or even annihilate it. Cancers often form but do not succeed well enough to out-compete our body's defenses and overcome our body's purpose to sustain health. In fact, being resilient and adept, but also quite selfish in their own right, our bodies are generally made to endure. As it turns out, cancers are likely coming and going all over our bodies throughout our lives. Perhaps even some nasty ones have formed, but our tissues can usually hold their ground and these cancers are eradicated. We are never aware of the cancerous transformations that are eradicated, of course. Cancers and cancer-causing cellular changes are therefore part of a normal life, but that means a great deal of our fate typically lies in how we deal with this enemy, and avoidance of toxins is not sufficient. Our power over cancer is not solely derived from lessening our exposure to the physical insults that induce it but by also taking steps to insure our cells are healthy.

Healthy tissues have more stable genomic material and therefore are less prone to malignant transformations themselves. Additionally, healthy tissues are woven with better materials and are therefore able to defend themselves against the cancers that are presented to them. It is not easy for a cancer to take hold of healthy tissue that has a robust genetic history providing for excellent defense, but, perhaps equally important, healthy tissue has its own desire to fulfill its destiny and is unwilling to change. In other words, we can resist a cancer invasion when our cells are in good working order.

Excess fat deposition in and of itself weakens normal cellular architecture, but obesity is a state that is arrived at by mechanisms that also promote cancer transformations and growth. In the over-weight state, anabolic hormones such as estrogen (obese men also have higher estrogen levels) and insulin are circulating, and glucose deposition escalates, making cell surfaces sticky and less functional. Inflammatory pathways also remain activated for no legitimate reason, and excess inflammation causes cellular chaos, even in distant sites, because cellular signaling becomes less functional with so much background noise as a result of the system's alert status. All of this is bad by itself, but these responses to obesity within the body contribute to two other very detrimental conditions as well: genomic instability and disordered energy balance.

Being perpetually dynamic, normal human tissues are constantly involved in the acquisition, utilization, and governance of energy. Corroded tissues do not transfer energy as they are intended to, and this weakness is a vulnerability. Cancers also need energy, and because they are entities that are growing faster than the surrounding tissues, they need more energy than most cells. So, they steal it. Malignant cells are the most feared because they tend to steal life from us, and weakened cells are no match for hungry malignant cells. This is why the person becomes thin and gaunt with advanced cancer. The normal body is broken down for the benefit of the cancer. Cancer begins as but a seed within us, but when our biological energy and tissues are corrupted and not put to good use, it blossoms perversely and flourishes.

As I have pointed out with regard to cancerous conditions, living within the Industrial Order includes *not* doing some things that would help us live a more healthy life, such as exercise, but we also do not eat high quality, highly nutritious foods in the Industrial Order. Lack of exercise could be viewed as a form of comfort in as much as one does not expend energy or deal with sore muscles and other outcomes, but the prospect of disease is a far more uncomfortable position. The sedentary life is associated not only with specific diseases and conditions but also contributes to premature aging in general. Indeed, if we act old we are more likely to feel old, frail and stiffened by joints frozen from disuse.

I once believed that calories are the same, no matter the source, but nothing could be further from the truth. I thought that all I had to do to be fitter, leaner, and healthier was to watch my calorie intake. This line of thinking is seriously flawed because not only are calories from manufactured foods bad in terms of their effect on us at a cellular level but eating bad foods deters the consumption of good calories, such as those from fruits and vegetables. That is, calories from unfavorable sources are not only corrupting our cells but also creating a health deficit as a result of missed opportunities to ingest high quality natural foods.

The chronic consumption of manufactured foods tends to change our very nature, making us less competent to benefit from the rest of nature around us and within us. By not eating foods that are natural, such as fruits and vegetables, we are detached from their inherently beneficial, mostly essential qualities. In fact, the absence of natural foods in our diet may do as much or more harm to our bodies as the presence of manufactured foods.

A diet that is deficient in natural foods tends to create a life of unnecessary, inappropriate inflammation affecting many organs, even the bowels. An increasing number of people in the US are suffering from chronic abdominal pain and bloody diarrhea as a result of inflammatory bowel diseases, such as Crohn's disease and ulcerative colitis. These conditions are receiving a great deal of attention these days due to the rise in incidence to a point where we routinely see the medication advertisements on TV and in magazines. Unfortunately, these conditions are the result of our dependence upon manufactured foods and the wide-spread use of antibiotics.

In the January 12, 2012 issue of the *New England Journal of Medicine*, Herbert Tilg, MD, writes of his findings that cruciferous vegetables (the Brassicaceae family), such as broccoli, cabbage, and Brussels sprouts, contain specific compounds that interact very favorably with our intestinal immune receptors and in effect provide better control of our intestinal immune system. The compounds found in broccoli improve the suppression of our immune system so that it will be less likely to attack our own body. Specifically, cruciferous vegetables have been shown to directly calm the rage of inflammation that denudes the intestinal mucosa and results in pain and bloody diarrhea. Therefore, it is what we don't do that results in auto-immunological attack on our intestines. But this is not the only means by which the absence of cruciferous vegetables harms us.

The microbiome of our intestinal tract is very much dependent upon the compounds in fruits and vegetables, and therefore a manufactured food diet alters the composition of this valuable asset (when it is functioning properly) because again it leads to the absence of essential foods for our optimal health. The fantastic eco-system within our intestines is only now becoming understood. This "magic carpet" of bacteria is part of our individual identity, and even identical twins do not share the exact same make-up of types and as regards the relative numbers of microorganisms. The selection for a different, less favorable intestinal biome is a result of two industrial-made insults. Manufactured foods and antibiotics select for a different intestinal biome. As we have recently discovered, the intestinal biome is extremely important to many body functions, and some of the less favorable conditions that result from its change include: promotion of obesity by abnormal function and regulation of gut hormones controlling hunger and digestion; altered immunity, and not only in the intestines as changes to the intestinal biome are thought to promote immunological attacks elsewhere in the body, leading to increased incidence of chronic arthritis, lupus, and other auto-immune diseases; vitamin deficiencies related to malabsorption.

Once lost, reestablishing a favorable intestinal biome takes time, and therefore the sporadic eating of fruits and vegetables is not nearly as beneficial as consistently eating them. The microbial population within us and on our skin also provides a reservoir of beneficial genes. In the *Scientific American*, June 2012 issue, Jennifer Ackerman reported that the intestinal biome is a reservoir of over 3 million genes (1,000 species of microorganisms) that code for the enzymes capable of digesting plant sugars. This is important because humans do not make the enzymes that efficiently digest plant sugars, and consequently the nutrients held within vegetables will likely remain bound in the plant material and pass through us. Enhanced digestion is a result of a better intestinal biome and this allows for better nutrient extraction from plants.

The direct effects of our diet are immense, whether from a cause and effect point of view as regards what we eat or from the indirect effect of what we are not eating. When we cut open an abdomen or chest during surgery, we can tell a lot about a person and how they live their life. I have performed hundreds of sigmoidoscopies (limited colonoscopies), and I can identify the people who ate a high quality diet and those who take care of themselves because it is apparent in their colons. The patients who eat a high fiber diet with less processed foods will exhibit pliable, glistening colonic tissue free of polyps, cancer, and diverticulosis. Even elderly people in this category possess colonic flesh that is supple, smooth, and outright glistening in the light. We do not need numbers from tests to evaluate our health; we can just look at our flesh. Invasive tests such as colonoscopies carry serious potential risks and can corrupt the most healthy flesh.

In the event of trauma, surgery, infections, or even cancer, our outcomes are largely determined by the condition of our tissues before the actual event or diagnosis. Inflammation and obesity tend to create the conditions for poor treatment outcomes because these conditions are associated with cellular chaos and tissue dissolution. When we live our lives separated from a biological purpose, we have lost our primary purpose.

MULTI-ORGAN FAILURE

When our bodies serve only the comforts and modern distractions of our minds, we can ultimately suffer from multi-organ failure. Total organism dysfunction is not difficult to identify, and we are generally in contact with examples in our daily activities. Many people you see cruising the malls or attending churches or schools will be experiencing multi-organ failure, and the manifestations are not necessarily subtle. I have noticed a growing trend of uniformity between the body shapes of the two sexes, for example. Men are growing breast tissue and wide hips while women are growing round protuberant bellies, often making their breasts appear smaller in relation to the rest of them. Functional capacity and endurance are also diminished in many among us, and sadly, this is occurring in young adult as well as older Americans (something I realized as I watched obese parents chasing after their children on a playground). Nature surely did not intend for men and women to have similar body shapes nor for obesity to become a disability epidemic.

The Disney animated movie *Wall-E* takes place in the future where machines serve mankind to such a degree that humans become extremely fat with short appendages and little physical distinction between the sexes. The cartoon was a satire on the current state of affairs, our modern society in which we are becoming more dependent upon our machines to support our lives. The movie gives us a view of our future selves as a population that has lost touch with our biology and the soil from which we arose as organisms.

In *Wall-E*, the characters' lives were given up for wanton desires and products of the artificial world, a message for us in this entertaining movie that is not so funny. However, given that multi-organ failure is an increasingly common condition of American life, it seems the joke is happening before our eyes. I will use diabetes as an illustration of how seemingly simple biochemical abnormalities, such as blood sugar levels, do not justly indicate the gravity of the situation for the patient and those people who are yet to be diabetic. Persistently high blood sugar levels make our cells sticky (like syrup), causing them to malfunction. When the cells fail, so do the organs to which they belong. Chronically high sugars tend to corrupt the function of pancreatic cells as well as the autonomic nerve cells, and consequently the pancreatic cells and neurons burn out, as do the blood vessels throughout the body. Diabetes damages the vessels in the eye and is the second leading cause of blindness in the US. The kidney functions to filter the bloodstream of toxins, but the high sugars associated with diabetes occlude the filters (capillary beds called glomeruli), causing kidney failure and becoming the leading cause of dialysis. The vascular beds are affected throughout the body, leading to high blood pressure, strokes, and heart disease. The destruction of the visceral or autonomic nervous system (as opposed to the sensory nervous system) makes successful compensation for low blood sugars very difficult, and over time the patient is at risk of excessively low blood sugars and even coma. As diabetes progresses, the sugars are more apt to routinely swing from extreme highs to extreme lows. A person with this condition has historically been referred to as a "brittle diabetic."

After roughly ten years, virtually all diabetics are expected to have what we call the end-organ complications of diabetes. One of the most common complications is diabetic-induced peripheral neuropathy, nerve irritation progressing to malfunction and eventually nerve death. In the first few years of the condition, there is numbness ascending the diabetic's leg and then it typically progresses to the transmission of constant burning pain. The patient's feet feel as if they are on fire, but they are in fact alerting the brain to the dire situation with regard to their metabolic derangements preceding the death of cells. For years, the nerves may relay pain, although nothing appears wrong to the naked eye. Some of our longest nerve cells extend from our backs to the ends of the toes, relaying information so we can walk, dance, run, and feel the earth beneath us. Eventually, the pain ends when the nerves are dead, but this indicates that the blood flow is diminishing as well, and the patient is susceptible to gangrene, often requiring an amputation of the foot or lower leg.

At first glance, diabetic neuropathy appears to be a situation in which the nerves in the feet are faulty and causing chronic pain, and physicians commonly prescribe anti-seizure medications to block the communication between the nerves in the feet and the brain. But a more accurate appraisal of the situation reveals that the original problem is far above the feet. The origin of this dysfunction is in the person's brain, not the feet or the pancreas (type 2 diabetes). The person's mind continues to eat the foods that cause and exacerbate his or her diabetes, so it is really more of a brain abnormality, and the other organs are simply collateral damage. Diabetic patients are generally sedentary people, and again we find that it is a problem of people *not* doing the activities that create healthy conditions for their entire bodies. Diabetes results in multi-organ failure, but so do high blood pressure and high cholesterol. Most of the chronic diseases are attributable to a common source: obesity and consumption of the modern American diet.

Industrial foods provide our brains with a feeling of comfort, and therefore healthy nerve function in our feet is secondary in importance. Generally speaking, the diabetic's feet do not provide discomfort to the brain until many years of recurrent cellular corruption have elapsed, but eventually the tissues in the feet react with pain after having been abused by the corrosive metabolic conditions set in motion by the patient's behavior. The patient's feet are simply crying out that the situation, what is going into the mouth and the TV time, needs to change. Our nerves have so much reserve built into their structure that they will endure several years of continuous abuse before they sound the death alerts, and by then it might be too late to change.

In fact, the insidious nature of our common maladies makes it all the more important to pay heed to lesser signs of dysfunction that are nevertheless significant and occur prior to any official diagnosis, such as shortness of breath when on the playground with one's children. In fact, for most people there will be signs long before the test numbers indicate a problem. The biological forces and organ dysfunction that are associated with most chronic diseases are already beginning the transformation of the patient into someone that they may not like to be, a sick person, perhaps disabled with multi-organ failure. If we are attuned to the biology of our life, our health habits, our physical performance, then we will easily identify cellular corruption inside us long before the doctor informs us. Attention to the obvious changes in our bodies provides opportunity for corrective measures before our organs are classified as failing.

All Americans should understand this process and the behaviors that lead to diabetes because everyone in America is at risk. As noted, some researchers suggest that by 2050 half of the US population will be diagnosed with diabetes. Many among the other half will have the disease and not be diagnosed or will have a pre-diabetes form of the metabolic circumstances. Diabetes may well become a common health state or even considered the norm, just as being overweight is now. Almost all of us can eat enough of the wrong foods to cause some degree of insulin resistance and other health afflictions. The American public is perhaps the most genetically diverse group of people in the modern world, and yet all Americans are at risk for obesity, insulin resistance, the metabolic syndrome and diabetes. That is because behavioral norms within the Industrial Order, such as calorie excess, can be attributed to the pronounced effects of manufactured foods on our brains, not that we have similar genetics. Thus all Americans are at risk of adopting the same behaviors that induce chronic medical conditions and multi-organ failure. Cellular corruption is obviously an equal opportunity endeavor in America, but just recognizing this fact gives one an advantage. If children are raised with the understanding that they are at risk if they follow the herd, quite possibly they will be able to prevent the suffering that the herd will inevitably endure.

It is unfortunate that the health industry continues to spend hundreds of millions of dollars per year looking for the common genetic abnormality that causes obesity when our population is so diverse. It is almost as if they are not looking at the actual people. The obese come in many genetic shapes, colors, and national heritages, which suggests that finding a common genetic cause appears to be a waste of money. The researchers also appear to have forgotten our recent history. Obesity was not an epidemic just fifty years ago, a further reason to suggest that the notion of genetic causes is farfetched. It seems an obvious assertion: every animal that takes in too many calories while trapped in a cubicle will gain weight, which is all the information we need to understand our obesity epidemic and to remedy the problem.

At the level of the individual, one does not need a medical degree to understand that signs of deteriorating health can be found in a tight fitting pair of jeans or the holiday pictures that contain increasingly rotund bodies, round faces, and double chins. Tools such as a bathroom scale or mirror can alert us to our increasing chances of chronic disease, and more direct evidence of the disease being established inside us can be found in the difficulty in climbing stairs. These are the indicators of our rising risk for cancer, strokes, diabetes, heart attacks, kidney failure, and other forms of suffering.

Cellular corruption and the prospect for multi-organ failure is not a medical mystery but a simple reality for many of us who frequently indulge in the all-you-can-eat buffets, vending machines, supersize chip bags, and fast foods delivered at drive-through windows. Unfortunately, the potency of our manufactured foods in corrupting our organ systems is so great that we need not require large doses to suffer serious consequences. The deleterious effects of manufactured foods occurs over time with consistency of the exposures. This will become more apparent as I discuss the subtleties of missed opportunities for health. I know that I will become obese if I eat the same foods that most Americans eat, such as the majority of the contents at the local grocery store.

Self-neglect causes the majority of American suffering, and the obverse is what we need in order to avoid much of that suffering — self-care and honesty. We are better problem solvers when we see that our problems are largely from what we don't do because only then can we ask: why don't I exercise sufficiently or eat a healthier diet? There will be reasons, but usually our answers will be traced back to our intentions for life, and if we misunderstand our basic biological needs that should drive those intentions, then our answers will inevitably be rationalizations. Health must be linked to why we live our lives or we are unlikely to sustain it, even with the best medical care available.

Yo-yo dieting and compartmentalizing unhealthy behaviors suggest intentions for life. Our fates for health can change for the better as we change our actions for the better. Excellent performance requires consistency throughout our bodies, from our minds to our cells — and the reverse, as we shall see.

We shape our bodies and our health just as we shape our lives in general, through our actions. We can change our lives (for better or for worse) because we are malleable and adaptable organisms. Natural tissue that sustains our life is never static, but instead an energized, fluid substance that can be manipulated in the moment, but capable of creating the enduring memories of each and every performance.

Our performance is the universal code of conduct in nature; it is what all life is about. If we are not physically performing as nature intends, then we must each ask ourselves: what is preventing me from fulfilling my intentions to live a healthy life? As noted, it is essential to understand cellular corruption, how it diminishes our physical performance and ultimately our satisfaction in life, but we must also understand our intentions, those that are a product of living within the Industrial Order. Our intentions are products of the mind, and they can be innate biological intentions, as nature intended, or influenced by cultural programming, and therefore we must be vigilant and resilient concerning the forces around us that would hijack our intentions for life and corrupt our minds, then our bodies, even at the genetic level. When the purpose of our life (the purpose of our cells) is not aligned with health, we are likely to be unhealthy.

We need not be trapped into being people we do not wish to be. Nature intends for us to be healthy, and although the Industrial Order may have other intentions for us, the choice between suffering or living out our biological destiny, true to a healthy life, is quite easy once one cuts through all the Industrial Order obfuscation.

CORRUPTION OF THE MIND

"Society never advances. . .
For everything that is given something is taken. Society
acquires new arts, and loses old instincts."
— Ralph Waldo Emerson

I asked my son, who is seventeen years old, "What is the purpose of your mind?" His response was succinct: "To serve the body." I thought his answer extremely accurate but also telling for the naïve child of seventeen knows an answer that seems more difficult to arrive at for an older person. It appears that we become corrupted as we age, forgetting basic things in life, like the fact that the mind is meant to provide for one's optimal health. This is rather interesting because we are supposed to know more about everything, including ourselves, not only as we age but because we are modern humans who have access to all the knowledge of our current age and the ages that came before. But the average American person's lifestyle does not reflect the truth that was evident within my son's response, does not reflect what far more so-called primitive peoples seem to know. Indeed, observing the day-to-day business of American culture, it is easy to conclude that there is an obvious disconnect between the intended purpose of one's nervous system, to coordinate the body's tissues to achieve the best health of the organism, and the goals one pursues during an average day.

The nervous system's control and coordination of our tissues occurs all the way to the cellular level, and there is virtually no location in the entire body that cannot be either directly or indirectly affected by the nervous system. Therefore, whatever governs the nervous system governs the body, and whatever forces shape the energy expenditures of one's nervous system will ultimately shape the energy expenditures of the body. In short, our bodies, indeed our lives, are shaped by the nervous system, and not figuratively but literally. The brain, the controller of the rest of the nervous system but also distinct in its processing abilities, is the ruler of our identity. What the brain believes to be real is as real as our lives can become. Although each one of us has a brain created with the intended purpose of securing the best health possible under all conditions, which is our natural identity and the innate intention for our consciousness, the brain can also hold values contrary to this primary directive.

At the point in my life when I felt I had become someone I did not want to be, I was fat and not well-conditioned physically, not quite ill, as in requiring medical care yet, but sick enough to realize that I needed to do something about my physical condition. My brain was not serving my body very well, and thus many of my daily choices were harmful. The cellular corruption that I had brought into my life by those choices had made me into a different person, both physically and mentally. The integrity of my tissues was less than ideal, my true identity being cloaked as my tissues were injected with fatty deposits, corrupting both my function and my form, my human competency, if you will. My vascular system was not toned nor optimally responsive, my muscles being somewhat atrophied and my joints carrying too much weight. But the most corrupted organ of all was my brain in as much as it was not performing its job: to secure the best health possible, the most optimal state to function in life. My mind had allowed my well-being to take a backseat to other concerns in my life. I was distracted by my busy life, of course, but the real reason was that health was not a priority in my life.

The brain is the organizational hub where sensory input is processed, its primary purpose to provide coherence to our existence and a coordinated response to all circumstances in life. The most primitive nervous systems, that of simple organisms, have one purpose and that is to move ingested materials through the body. These creatures do not have brains, but their nerve networks make their existence more efficient. As we ascend the ladder of organismal complexity, there are benefits to a dedicated processing system that coordinates many organs and tissues into a common purpose. Within the more complex systems of mammals, goal-directed behavior arises and life takes on additional color as the organism functions beyond just reflexes. The human brain stands as the ultimate processing center, one that not only calculates and reflexively responds but contemplates, pursues desires, and weaves instincts into newly contrived knowledge, perhaps ultimately producing wisdom. No computer will achieve wisdom as does the human brain, nor provide the perspective we need to live the best life possible. Logic is different from wisdom.

But despite the virtue of having the largest brains on the planet, despite being the creatures capable of wisdom, most humans' brains in the developed world are corrupted. They are corrupted because we tend to forget that the primary concern of a brain is for the care of the body within which it resides. The modern human mind seems to plot our days, tending to the urgent concerns of the Industrial Order, and allow the body to be torn down needlessly. Worse, a mind corrupted by the Industrial Order can be convinced that the body within which it resides is healthy, even if that is far from the truth.

The Industrial Order will bend the truth, if not lie outright, to serve its own needs, and it seems this corruption of our sensibilities extends beyond questions of the marketplace and how those who benefit from our corruption manipulate us with advertising. Our notion of ourselves includes an idea that our government is supposed to be more attuned to the real needs of its citizens and protect us from profiteers who would abuse us in one fashion or another, the government serving the common good. Indeed, the US government indicates that our health matters with the institution of regulations regarding food safety and the like, but the same government creates the conditions wherein candy bars, alcoholic beverages, pastries, corn chips, Whoppers, and soda pop are less expensive sources of calories than apples, pears, grapefruits, and other natural foods. The US government subsidizes grain and corn production, which in turn enables the manufacture of cheap artificial foods. In this fashion, the US government distorts our food reality by economic means. What is even more concerning is that our food tends to influence our health directly, and therefore, by altering our food realities, the US government is also changing our health reality at the very same time. Through economic incentives, we are coaxed into consuming cheaper calories that are delivered to us in non-nutritive foods and therefore prodded into being an unhealthy people.

Many public figures put forth the argument that our choices are important in our matters of health, and I firmly believe in our right to personal choices. However, our choices are limited by extrinsic factors over which we have no control, factors that limit the choices we have and that influence the choices we make. For instance, if the availability or cost of two choices is very different, then which option we choose is being greatly influenced. We see this with foods and medicines. Manufactured foods are vastly cheaper in comparison to whole foods, especially compared calorie for calorie. Medical care is generally cheaper for an individual through government and employer subsidies than through fitness equipment or time away from work to engage in self-care.

Furthermore, intrinsic factors can alter our choices greatly. The delivery method for 100 calories of energy differs between an apple and Doritos. Most people understand that a very smart monkey will "choose" cocaine mixed in water over food, even though the animal will subsequently die of hunger. This monkey had a choice, some people might say, but the deadly choice has factors in its favor that exceed common sense. Indeed, free will is a tricky concept in any scenario, but when mankind can make things, can even fashion our sense of free will, the trickiness grows exponentially.

To what extent are we exercising free will, for example, if the information regarding the choices is grossly distorted unto being an outright lie? Marketing campaigns can consist of lies and still be so pervasive that we lose our sense of reality and accept these false claims. More often than not, in fact, American products are not represented as they truly are. This is done on the grandest of industrial scales, and our biology is not equipped to handle a threat of such magnitude. We can only contemplate truly free choices if we have the correct data, and I would argue that otherwise we are not really having a free will.

The unhealthy consequences of this state of affairs are insidious. Millions are content on the sofa, consuming artificial foods and watching the advertisements of medical rescues and pharmaceutical remedies parading across their TV screens. With the acquisition of the industrial identity, our brains have a difficult time discerning that all of this can be quite harmful, and in fact our notion of our lifestyles is that they are sophisticated. These transformations in the mind become transformations of our bodies.

Through my extensive interactions with patients over decades, I have realized that many Americans lose mental competency in health matters, that they tend to make choices that are bad for their health; but these choices are spoken of as though the patient in fact had very little choice in these matters. We know that peer pressure can produce results like this, people signing on to actions that they would not have otherwise — and three hundred million people is a lot of peer pressure. Of course, a quest for comfort drives a great deal of our behavior, but there is also a very important problem usually overlooked that affects our choices with regard to health. Within the Industrial Order, people tend to lose competency when their health problems surface. This loss of health competency is not only a product of their routine choices that corrupted their physical tissues, but the loss of tissue integrity further limits their choices. It's more difficult to contemplate running after a heart attack or after one achieves morbid obesity.

However, some of our so-called choices are masked by a reality which is difficult if not impossible to discern, and such is the case with our food choices within the Industrial Order. My father used to say when I was a child, "Don't eat corn because farmers use it to fatten pigs." I distinctly remember him saying these words as we sat around the kitchen table with my grandparents and the bowl of corn was being passed around. What is interesting is that my father consumed huge quantities of corn over the course of his life in the form of other products. In fact, to this day he eats a great deal of corn, but at eighty-plus years old, he has a difficult time believing that the products he consumes, which do not appear to be corn or have corn in them, are actually composite structures of corn. Tampico fruit juice, salad dressings, corn chips, ketchup and other assorted American condiments and foodstuffs don't look at all like corn but are to large degree.

The American omnivore certainly has a dilemma, as Michael Pollan described rather eloquently in his 2006 treatise. Pollan, with assistance from a biology scientist at Berkeley, brought to our attention that unique carbon isotopes from corn, as analyzed by a mass spectrometer, are found in many items in our food chain that do not resemble corn. Foods such as French fries, soda pop, even hamburgers tend to consist of carbon atoms from corn. Perhaps if we studied my dad's tissues or other Americans' tissues, we would find that we are largely made from corn atoms.

More sinister than the fact that my father (along with the rest of us) was duped all of these years is that the corn products made by industrial might are much more harmful to us all than eating the corn kernels used to fattens pigs. I consider my father, who is a post-graduate fellow in Foreign Service from Georgetown University, to be one of the most discerning people I have ever known, but he is unable to discern extremely important aspects of his life, like the truth about his food, given that he is a product of the Industrial Order wherein reality can be most unreal.

With ingenious technological finesse, our inventions change our perceptions of real food and health so subtly, so seemingly innocuously, that we are usually not aware that it is happening at all. Industries can corrupt even the most educated minds because industry can change a culture by virtue of the products it delivers. Cars are delivered to us, and therefore we need not walk. Personal data accessories (PDAs) are omnipresent, and therefore we need not remember as much (and because what we do not use tends to atrophy, our memories as well as our bodies are being lost to these innovations). Smart-phone technology is a premier example of a ubiquitous industrial machine sold as a tool of personal empowerment, but in truth the smart phone can reduce our independence and likely alter our brain's anatomy by changing our patterns of thought and resulting behaviors. The smart-phone device typically becomes so aligned to one's life, unto becoming one's social essence, that it becomes literally a body appendage. I see this happening in my own children — and our youngest citizens are the most susceptible to an enduring device relationship — and so the species is moving closer to the loss of not just face-to-face communication but to a good part of our capacity for memory and other brain functions.

A brain is intended to coordinate a coherent reaction to the circumstances around the organism. It does so by using the available information from its senses to secure the best life possible. For most organisms the best life possible means the best health possible, but the human brain, by way of the mind, frequently has other intentions in life aside from its primary purpose of survival and optimal function. Many of these other intentions and aspirations lying outside of the purview of health are dictated by society, and as the priorities of the mind drifts from its primary purpose, coordination of the body's tissues into the most optimal physical condition, the mind is in effect becoming corrupted. Thus our enmeshment into the Industrial Order is a position of great risk, one that can affect our body by diminishing our life skills as individual organisms.

One's mental competency regarding one's health, or life skills, is generally reflected in one's tissues because whatever influences the mind will invariably influence the body. This is a very holistic way of looking at not just overall health but chronic disease, because it is the loss of self-care skills that leads to the prime conditions for a chronic disease. Once the chronic disease state is reached, sickness debases one's life skills further, and one's reality and the mind's capacity to deliver life skills change for the worse, a spiral that has become much too "normal." Within this all too common spiral of suffering attributed to diminishing life skills, one's ability to accurately discern the cause of, and remedy for, maladies becomes increasingly less potent. We are made less competent as organisms.

Improving our skills to provide self-care, the healthy behaviors that assure a better survival biologically, can prevent disease, ameliorate disease, or even cure disease. Our success in life, as measured in Industrial Order terms, requires very few survival skills and even less physical function, however, and so it is no wonder that our priorities shift to pleasures that do not kill us immediately but can eventually. Our genes are not equipped for Industrial Order life, and therefore, not only in spite of but because of the Industrial Order's tendency to diminish this capacity in us as regards our health, we must cultivate the most quintessential discernment to survive our culture.

Precise discernment is mankind's pride and joy in many ways. We can compare the intricacies of apples to oranges, calculate future results for myriad things, predict seasons and even weather patterns with some accuracy, and we make conclusions about the good and the bad and then convince other people of our beliefs. We can do many things with our mental acuity, and discernment is the crux of them all, and it is therefore most intriguing that, despite that prideful feeling, perhaps a self-righteous feeling, derived from our capacity for discernment, we appear to be losing that very skill, the very skill that separates us most from the animals and makes us most human, in regard to our health.

Entering the medical system is not likely to bring a realization that we lack the basic life skills to be healthy, the primary cause of most of our infirmities, let alone give us access to those life skills. Modern healthcare does not enhance our power of discrimination or discernment in matters of health — in fact, quite the contrary. In my interactions with patients, I see how they are confused about their physical health because of the Industrial Order within which they live and from which they then enter the medical arena, only to be confused all the more. The Industrial Order provides each one of us with a grandiose entitlement toward technology, for example, and that sense of entitlement is a fire made greater with the technological fuel of the medical centers. Americans deeply believe that we are not only entitled to any technology available but also assume that the technology available will automatically be good for us.

We not only assume that technology will give us what we need when it comes to our health (and all other aspects of life, for that matter), but much of our perceived needs are dictated by the presence of the technology itself. For instance, over my entire career as a physician, until very recently, I recommended PSA tests to prevent the negative effects of prostate cancer; but now we are to stop ordering screening PSA tests because, after all of the assumed benefits, the reality was different. The PSA test hurts more people than it helps because most men with prostate cancer do not need to be treated (the cancer not being significant to their lifespan or quality of life). Several hundred million PSA tests were run on men over several decades, and all the while people assumed men needed them, but the invention of the PSA test itself seems to have driven the need for more tests. Screening for prostate cancer is a noble idea, and the PSA was elevated in many men who underwent prostate biopsies, prostate resections, and prostate radiations. But we have confused a noble idea (screening for cancer) with the need for a test that is severely flawed. Nobody paused long enough to discern whether or not the cancer that the PSA was measuring ever required treatment — in other words, whether or not the PSA was a good screening test. The PSA test, like many screening tests, represents more of a pure desire to squelch our mortal fears than an actual need for a good and healthy life.

This is similar to the perceived need people have for the use of antibiotics for the common cold, pharyngitis, bronchitis, sinusitis, and ear infections. Antibiotic technology actually drives the need for antibiotic products independent of one's true need. Many times a day in any primary care doctor's office, physicians battle with patients over the need for antibiotics, because patients have technology entitlement issues related to the Industrial Order's training throughout their lives. It seems that our instruments of precision, those tests and procedures and technological interventions, don't necessarily provide precision at all let alone our Industrial Order expectations of precision.

In fact, when a physician withholds therapies or testing, the patient suspects the physician simply does not care rather than that the doctor has determined those therapies and testing will do the patients no good — or even harm. The presence of any technology seems to dictate that it shall be used, one way or another, regardless of absolute need, and our actual needs for technology are more and more obscured as our fears and desires set in. The fact that malpractice lawsuits (another one of mankind's inventions) exist impacts the need for technology too, because physicians are far more likely to be sued for under-use of technology than over-use.

Obviously, assumptions that technology will heal us are a poor substitute for a truly healthy reality built out of how we live. But in spite of all the evidence that Industrial Order living is unhealthy and that entering the medical-industrial complex will not restore us, our assumptions persist. These assumptions are translated into actions and behaviors that cause us to be overweight and unwell in the first place, then over-tested and over-treated; and eventually our assumptions regarding technology change our biology, reaching the core of our existence, our genome. Medical centers teach us skills such as a pill science, pill regimen, procedure schedules, and other "innovative" lifestyles, but these are typically more productive for the financial bottom line of said medical centers than they are life skills we lack. Modern healthcare can be dangerous in large part because it compounds our Industrial Order assumptions, such as all food is nutritious and all healthcare is good.

It is perhaps telling that my medical education did not make me more competent at sustaining optimal health. The competency one needs for optimal health is not learned in classrooms or from medical centers nor indeed found within the medical-industrial complex. You would think, to watch the species, that humans had cornered the market on health with technological interventions, but the many other healthy organisms in this world, organisms outside the Industrial Order, would suggest otherwise.

We are obviously helped by many aspects of the Industrial Order, but this automatically makes us less likely to see its negative aspects, and we are therefore likely to be a bit soft on its assessment for harm. This success bias is innate, actually woven into our DNA as we have been imprinted and trained rigorously within the tenets of the Industrial Order, making it quite tolerable and extremely difficult to dismiss. Such a prejudice is expected, in other words, and therefore I know that many readers will deem the principles I have laid out a conspiracy theory of some sort, a theory that the Industrial Order is controlling us and the medical-industrial complex deeply implicated. However, a conspiracy theory suggests that this is all directed, a plot, but in fact this is merely where our cultural evolution, especially our reliance on technology — in concert with capitalism, of course — has brought us. The manufactured food and medical industries merely operate in a symbiotic fashion. But it need not be this way. The effects of this Industrial Order identity I am expected to have is to give away my health to spasms of technology within the time span of pixel refreshment and without ever breaking my gaze from a TV or computer screen. I am as vulnerable as anyone to the prospect of accepting the tenets of the Industrial Order as the most important aspect of reality, but we can defy that order by merely re-realizing our biological identity.

The first step might well be a critical analysis of our Industrial Order lives. Sometimes, if we step back and look at something, we can see that, regardless of the improvement in our lives we receive from a given entity, the profound magnitude of its influence upon our lives also contains an inherent potential for harm. Many naturalists have found this stepping-back move easier by first considering less complex forms of life and extrapolating meaning to the circumstances within our complex civilization. For example, Harvard professor and author E.O. Wilson reports a particular interest in ants and ant antics since his early childhood, but his professorial studies of those tiny creatures have provided great insight into the workings of our human colonies. A like analogy may be helpful in illustrating the degree of occult influence that the Industrial Order has upon our lives.

A tree is a rather simple specimen of biology compared to mammals. The cells of trees are routinely studied in high school biology classes, wherein children first encounter the unseen, microscopic world. Tree tissues are divided into the xylem and the phloem and specialized cells. Chlorophyll and its role in photosynthesis are reviewed all the way to the electron transport chain that produces oxygen for us to breathe. The botanist can describe many of these details as they peer into a light microscope, later into an electron microscope, and then finally into the biochemical equations explaining the actual molecular reactions that are associated with tree life.

The tree has an extremely important challenge that actually defines its entire existence, but that challenge cannot be realized by our scientific descriptions of its cell structure and metabolic activity. A tree spends its entire life in defiance of the unseen force of gravity, and it is this challenge, this relationship to planet Earth, that makes the tree an authentic tree. If we take the tree cells to outer space, the tree will not thrive or even grow as an authentic tree because the gravity is gone. Upon failure, we may peer under the microscope to assure ourselves that all of the organelle contents are present and accounted for, that the water, soil, air and atmospheric pressure is just right, and yet the these elements did not grow to be a real tree because the whole process lacks gravity. Finally, we may determine that the presence of gravity is lacking, but we can never know how the tree relates to gravity or how its cells respond to the struggle as a defining force for its life. We may draw substantial conclusions, but they will not explain the lack of a tree from the other elements we deem necessary to treeness.

In fact, one could argue that gravity is a detrimental force against the tree, always pulling it down and not building it up. Gravity just is, and a tree simply needs it to live, but we do not need to understand everything to benefit from the relationship. Conversely, the Industrial Order can appear to always be building us up but meanwhile be tearing us down and tearing us apart by increasing the entropy of our lives, yet remaining forever indescribable and resistant to discernment. The influence of the Industrial Order cannot be distinctly measured for good or for bad; it just is a fact of our lives and we must constantly negotiate with it as the tree negotiates its life in the constant presence of gravity.

Just as we cannot completely understand the tree's life if we study a tree outside the context of gravity, similarly, if we study the average American's health out of the context of the Industrial Order, we will be oblivious to the "gravity" (all puns intended) of our controlled circumstances. The influence of technology and industry on our lives is multi-dimensional, affecting our thoughts, our biological fabric, and eventually the spirit of our lives. As we grow into the Industrial Order, we are enmeshed within it and infused with its products. We in effect find it difficult to be separated from it, even at the cellular level, by virtue of the foods and medicines we ingest. All that I have described above may sound like something from a science fiction movie, a fantastic paranoia or pronounced craziness, and I grant the reader such a consideration; but before conclusions are to be drawn about my creativity or sanity, one should review humankind's recent history with a more discerning eye. By doing so, one shall see that the presence of the Industrial Order within us is not only plausible but rather predictable.

Society changed rather dramatically when mankind adopted grains and agriculture for survival, abandoning the hunter-gatherer lifestyle as perhaps too unpredictable, too variable. As mankind became dependent upon grains for sustenance, the grains then cultivated man's societies, so to speak. Bread not only changed man's relationship with food but symbolized man's supremacy over the natural world. Likewise, industrial food production has given food new qualities and even a new definition because it need not be nutritive and yet can still be considered as food, and indeed, our manufactured food products now manufacture our society. As our food became corrupted, so did our minds. As we became more technology dependent over the last fifty or so years, we witnessed the medical-industrial machine gain prominence over much of our lives, and that machine is now poised to consume the majority of revenues generated from our labor, in part by virtue of the corruption of our food.

The medical machine has sold us on a new definition of health that is perhaps most convenient to the industry itself as our medical technology itself breeds new needs and desires for life within us. Just as industrial food altered our relationship with food, industrial healthcare has changed our relationship with health. The best customers are the dependable ones, of course, and therefore it is to a company's advantage if its customers become dependent. Industrial medicine thus treats patients as what they are in any for-profit environment: customers. No one ever stopped to think that dependence on industry may not be healthy in and of itself. Potent but occult forces (as in hidden from view) act directly on our minds and shape our lives like gravity shapes the life of the tree; but whereas the tree need not question the effects of gravity, for it enables to the tree to be a tree, we should question all of the effects of the Industrial Order, even those that emanate from so-called trusted sources, because the obverse is true of our relationship to this force in our lives — it is making us something less than human. For example, the food and medical industries are changing our relationship with our biological selves, to our individual and collective detriment.

FOOD OF MAN

During the Paleolithic, which spanned 100,000 years or so, man was nomadic and lived off killed meat and gathered fruits, vegetables, and nuts. During this age when our ancestors were hunter-gatherers, mankind underwent the most change in his brain's evolution. The average brain volume grew from that of a chimpanzee (850cc) to the standard issue 1200–1300cc brain humans have today. It was this increase in cerebral capacity that led to sufficient intellect to create civilizations. Once humans were endowed with a mind that could create, the creation of food was in order.

Bread has the distinction of being the first manufactured food. The early Neolithic people must have felt a deep satisfaction as they could live off their own food creations. These Neolithic people must have also sensed a certain power in this new ability, perhaps even an elevated importance in the universe because they now had the ability to be an active part of the food production process rather than just passive consumers. They doubtless had no idea of the power they were only beginning to unleash. Within a span of only 10–12,000 years, a mere flash of geologic time, one species, mankind, would alter the entire biosphere.

Within the Fertile Crescent of Mesopotamia, early civilizations cultivated emmer and einkorn wheat to make their breads; these grains were very difficult to grow, and so technology was necessary to coax the production of these grains on increasingly larger scales. For the first time, an organism on planet Earth was meticulously controlling natural resources like water, soil, and food grains to guarantee its survival. As this unfolded on an ever-increasing scale, man's effect on the planet also increased dramatically.

The advent of farming and agriculture that is the hallmark of the Neolithic Age also proved to be the root of civilization. Indeed, it is bread that most symbolizes man's conversion from a wild animal to a civilized being. The newly established towns, and later cities, that this new agrarian lifestyle allowed gave rise to written language. These larger aggregations of humans also led to religious sects and more complex governments. Writing allowed our thoughts and beliefs to be transmitted in a consistent manner over time, and history could be recorded.

In other words, this new agricultural lifestyle produced more than just grain products; it also produced a new path for the evolution of our species. Grains changed our behaviors, our values, and even our physical statures in a very short burst of time. Grains were made into breads, then alcohol, both altering the future of man. They were placed in ancient graves to feed us in the afterlife. Symbolically, the breaking of bread solidified order in great societies, and thus the importance of manufacturing bread was elevated. Food became a sign of success and luxury, not simply an energy and nutrient source. Celebrations were enhanced with pastries and alcohol. Mankind's success at bread manufacturing led to more manufactured foods and the cultivation of other foods.

Fast forward several thousand years, and grains are still altering the course of humankind. Our relationship to food over the last few hundred years increased exponentially in the direction of food no longer being used for just nourishment, this change more pronounced as food became increasingly manufactured, increasingly a manmade object to be used as we saw fit. Through the development of a grain-based diet, food has in fact become a drug, a powerful weapon or coercion device, and a symbol of our success. That is, pastries and other grain-based delights have historically altered our brain's neurochemistry, much like drugs, and even induced over-consumption. War changed as a result of these early farms because they had to be defended at all costs. In developing countries, grains are still being used to coerce and control the masses, much of this control of the food supply related to ethnic tensions. These attributes are far removed from food simply being a nutrient source. Like all commodities, we now manipulate our food on an industrial scale. Food has not only become a complex creation through scientific means but is now so abundant that we see fit to waste a great deal of it, by some estimates 30–40%. But most important, as has already been discussed, there are substantial health consequences, both physical and mental, to the changed relationship we have with nutrients.

Obviously, I have intellectual curiosity regarding our history in relation to grains because I find it fascinating how we have altered our culture and behaviors as a result of this early technological achievement. As we can now see clearly, the advent of bread and the advent of civilization, including industrial foods, are linked philosophically, biologically, and for some, even spiritually. So our ties to grains run deep, and anyone who proposes that we should reduce our bread consumption is likely to be met with skepticism and possibly even viewed as an uncivilized person. As it turns out, our emotional connection to bread is not simply an ancestral phenomenon, however, but also a function of neurochemical effects on our brains caused by grain consumption in the form of manufactured foods.

Eating any food is a visceral experience because of the relationship of food to our survival as well as the effects on our brains, and therefore the relationships we develop involving food are likely to affect us in a deep way. This makes changing what we eat a difficult task, no matter the food source let alone the added difficulty represented by changes to our brain caused by manufactured foods and all that manipulative (and misleading) advertising. My personal relationship with food changed only after I had an epiphany about my physical condition and earnestly wanted better health. Once I determined food to be important to my health, and not just quantity but the source of that food, I decided to learn more about what I was eating. The dangers of high-fructose corn syrup and the glycemic index were then in the news, and, in combination with reading Michael Pollan's book *Omnivore's*

Dilemma, I realized the need for more scrutiny of the manufactured products I consumed. In order to avoid high-fructose corn syrup, I had to start reading labels, and this meant I had to stop buying certain products.

Our family typically ate pancakes on Sundays, and so we also ate syrup. I decided to purchase real maple syrup for the first time in my life, and I was amazed how difficult it was to find in the grocery store. When I did find some, I was shocked by the price, which was outrageous, but I bought it anyway, for the sake of our health. Interestingly, when I tasted real maple syrup for the first time, a strong unease went through me because I realized I had been deceived my entire life. I had never ingested real maple syrup, despite believing I was sweetening my pancakes with it. What I had believed true was a fantasy created by the food industry.

The food companies had altered my maple syrup reality, in other words, and in fact I did not like real maple syrup as much as the high-fructose corn-syrup version. This was my first realization of the power that industries have over our minds if they could thus alter reality, make me believe this fake stuff was not only the real thing but that it was actually better than the real thing. The maple syrup example is not simply about the purity of truth, but emblematic of the overall issue and thus instances of physical suffering when we are not aware of our relationships with various industries in a position to negatively affect our health.

Incredibly, I was forty-one years of age when I realized that Aunt Jemima had been deceiving me all of my life, a realization that prompted me to research the other products I was consuming. When I discovered that Fruit Loops have no fruit in them (none), OJ is not necessarily made from oranges, and blueberry pancakes in the frozen section have no blueberries, I felt somewhat betrayed by society and even my parents. How could they let me believe that stuff?

I have since become a more sophisticated consumer, and as a result, I have often wondered why I placed so much faith in the manufactured food industry in the first place. Even as an educated adult, I spent decades disbelieving the people who said all of the artificial ingredients in our food supply were bad for us. It was not important to me that chili-flavored Fritos consisted not of actual chili but chemicals that led my brain to believe my tongue tasted chili. The fact these foods were constructed in factories never bothered me or gave me pause, not once, before I shoved them in my mouth repeatedly.

Not so long ago, in fact only prior to my parent's generation, food was expected to be provided by nature, not factories. I remember my grandparents commenting on how they could not believe that I would choose foods from factories over food from the ground. Nana would say, "Why, honey, you don't know what you're missing with these green beans." There was far more truth in those words than I knew at the time, or indeed that she could have intended, not knowing as I do now the cellular benefits of what those beans contain. I remember that my grandparents not only had a much stronger affinity for food from the Earth than I but a subtle distrust of the foods in a bag. They certainly did not latch on to the newest food creations as I did, and I thought it was it was just because they were old fogies. I see now that my grandparents had a different relationship with food than I did. They ate for real nutrition and for natural taste, whereas I chose food based on convenience, artificial taste and many nebulous attributes I could not put my finger on, like those the ads conveyed or the changes taking place in my brain because of the chemical reactions these foods engendered. As it turned out, the calories I consumed were empty, only my assumption of nutrition was real.

A good business plan must keep consumers participating, and in order to stay ahead of the curve and keep us engaged for revenue's sake, the food industry hires brilliant scientists so that it can constantly adapt to the public's tastes, even to our concerns about the healthiness of their products. When there are health concerns about a product, the industry has ways to minimize our concerns. For example, "healthy choice" entrees have suddenly appeared on menus to bait consumers into the restaurants when they would otherwise be avoiding them for health reasons. Patrons may rationalize going into the fast food establishment with "healthy choices," but, once inside, the familiar aromas of French fries and hamburgers entice them on an emotional level.

It is not difficult to understand how it is that the effects of manufactured foods (consistently designed with high sugar and high fat content) are infinitely more reinforcing to a consumer's behavior than orange slices or apple wedges. Our concerns regarding nutrition are overwhelmed when neurochemical effects entice our brain's reward system, and our resolve to choose the most healthy selections dissolve. "One more meal will not hurt me," I used to say, until I realized that the effects of these modified molecules on our health are immediate as well as cumulative.

Another example of marketing finesse is a mainstay of the manufactured food industry: the transformation of blatantly unhealthy products into "healthy products." Chips are being baked instead of fried, for example, but they are still very unhealthy. In a very inventive fashion, we will find artificially made products containing specified vitamins and minerals so they can be advertised as healthy, or at least healthier than the previous food model. Of course, they are not any healthier, and the devil is in the details. For instance, products are allowed to say they are trans-fat free and show zero trans-fats on the nutrition label as long as they contain less than 0.5 milligram of trans-fat per serving. This may seem reasonable, but serving recommendations, being generally insufficient to satisfy the average American's palate, are never heeded. So, with the consumption of more than one serving at a time, and then adding those multiple servings over the course of a day, we have a new problem with our labels. The zero classification that does not really mean zero causes real health threats. In fact, it has been shown that only 2–4mg trans-fat per day is harmful enough to cause blocked arteries or atherosclerosis, and thus a harmful dose of trans-fat can be found in eating only four servings of the "zero trans-fat" foods. Trans-fats were invented as a food additive and applied to the manufacturing process to increase the shelf-life of factory-made foods, and so the only way to actually avoid trans-fats and not receive the harmful dose is to avoid manufactured foods altogether. This is yet one more example of how the government manages to allow the food industry to circumvent meaningful regulation, regulation that would benefit our overall health, if zero really meant zero.

However, as noted above, the US government is complicit in the current control that the food industry has over us in many more overt ways. For example, most manufactured food products are reconstructed corn, wheat, and soy beans. These are heavily processed foods; the atomic signatures of their constituent grains are all that remain of those actual grains after processing and being fashioned into what are ultimately artificial products. These particular grains are so universally used for food manufacturing purposes because they are so cheap. The US government subsidizes their production and so the cost is not determined by real free-market forces. If these products were allowed to cost what they should, then industry would be less motivated to fashion these crops into manufactured foodstuffs. However, so-called cheap foods really cost us more in the long run because the costs are made up in medical care for the unfortunate consumers. Therefore, the US government is simply a conduit or intermediary between the food and medical industries. Moreover, it seems a bad idea to leave "fixing" our health problems caused by our diets to the medical industry, which has a vested interest in the status quo and an unhealthy belief in technological intervention.

HIDDEN MUZZLES

People with morbid obesity suffer a great deal, and unfortunately the population of obese patients is expanding very rapidly in number. All cases of obesity result from too many calories ingested and too few calories expended, the same thing that is wrong with the simply overweight patient, and evidence-based medicine has invented some interesting innovations to fix the problem of too many calories in and too few calories out.

Surely my readers are aware of the many medications used historically for dieting, such as amphetamines and other brain-altering pharmaceuticals. These have never been proven to actually improve health over a lifetime, and yet they are still being used. Many of these drugs have substantial side effects, some of which are permanent, including heart valve abnormalities and addictions. New medications have just been approved for use by the FDA, including Qsymia, a drug combination consisting of the old amphetamine phentermine (part of the phen-phen craze) combined with an anti-seizure medication. Editorials are already appearing in medical publications criticizing the FDA approval as the drug has serious side effects, such as heart attacks (cardiovascular events ranged up to 300% risk). Michael Lauer, MD, wrote an excellent article, "Lemons for Obesity," in the 17 July, 2012 *New England Journal of Medicine,* and he sat on the FDA panel, voting against the new drug's approval. Dr. Lauer also provided additional, more holistic insight into concerns about these new medications in that they are allowed to come to market with an "illusion of validity," assuming ". . . just because a drug reduces weight and improves some biochemical markers that it will be safe, let alone effective." A person's general well-being should be considered much larger than a few biochemical surrogates, yet most consumers assume that changes in extremely small molecular measurements translate into a better overall health status. Use of surrogate markers of health is a great flaw in modern medicine because it narrows our vision in such a way that we are susceptible to chicanery, even the kind approved by the FDA.

Another medication just approved is Belviq, which acts on serotonin receptors, the ones involved in mood stabilization, and therefore it helps persons cope with their American lifestyle. Both of these new drugs are used in effect because people refuse to learn new life skills. Despite the fact that obesity and many chronic diseases are a result of a poor relationship with one's self and the fact that technology carries some of the blame, the need to apply innovations has not diminished. In fact, medications are not the only proposed treatments for obesity. Some physicians choose instead to view the obesity epidemic as an anatomical problem and think therefore that surgery is the best option.

Bariatric surgeons routinely disfigure peoples' anatomies to physically reduce the amount of food those persons can take in. These specialized surgeons enter a person's abdomen, generally using a scope, and rearrange the patient's stomach and intestinal anatomy in a fashion that induces a perpetual state of malabsorption so that the patient does not absorb or assimilate all of the calories he or she ingests. Bariatric surgeons therefore surgically alter a person's normal anatomy permanently, even though the person's problem was not anatomical in nature. This seems a perturbed notion of evidence-based medicine.

There are serious potential complications to these surgeries, such as life-threatening infections, depression, suicide, ulcers in the pouch that is created by surgical attachments, bacterial overgrowth in the intestines, and severe vitamin deficiencies. For the rest of the person's life, the absorption of extremely beneficial nutrients, such as antioxidants and vitamins, becomes nil. Essential vitamins and minerals such as vitamin B12, thiamine, folate, copper, zinc, and vitamin D are routinely defecated out since their absorption is dependent upon the optimal function of the intestines, which are now deranged. These patients are at risk of pernicious anemia and neuropathy (debilitatingly painful feet) due to vitamin B12 deficiency. As an internist, I frequently deal with the long-term complications of these surgeries, such as terrible bacterial infections and small bowel obstructions that require hospitalization and possibly surgical release of the adhesions. These bariatric surgeries induce significant scar and adhesion formation within the abdomen of the patient because they tend to tack down the bowels in such a way that, later, stool is not allowed to pass through, and so chronic abdominal pain is a very common problem after surgery.

Curiously, before being permitted to undergo this extraordinary intervention, the patient must sign a contract with the surgeon promising to attend classes about emotional eating, to get sufficient exercise, and to engage in healthy dietary interventions before and after the surgery. In fact, the bariatric surgeons say emphatically that, if the patient does not follow a strict diet plan and engage in meaningful exercise for the rest of his or her life, then the surgery will likely not work. The reason is that the human body is a fascinating piece of biology and can adapt to the new surgically created gut configuration, thereby allowing the patient to gain some or all of the weight back that they lose initially. The surgery is hardly a miracle, because if the patient remains in the same lifestyle habits as before the surgery, the results are largely nullified. People will still tend to be unhealthy if they eat factory-made foods, but smaller quantities. One wonders: why not skip the artificial intervention and go straight for a new lifestyle that will preclude the need for the intervention?

Because gastric-bypass surgery is extremely disfiguring and fraught with serious complications, bariatric surgeons have devised another less invasive method to restrict a patient's propensity for over-consumption of calories. Through a couple of scopes inserted through the patient's abdominal wall, the surgeon applies staples or rubber bands to make the stomach smaller, which is referred to as the Lap-Band procedure. It does not require a significant amount of understanding to see that this is indeed a sophisticated muzzle that is only hidden within the patient, placed intentionally to limit his or her food intake.

Like the more invasive surgery described above, these patients must also engage in a healthy lifestyle or the band does not work. Interestingly, if the initial Lap-Band procedure fails to produce the desired weight loss, then the surgeon goes back inside the patient and adjusts the band a little tighter. Of course, there are limits on how frequently this can be done, only every several months. It would almost be comical, if it were not for the fact that these patients are suffering, that many patients come into my office after the Lap-Band and state that their bariatric surgeons will not adjust their gastric bands tighter anymore this year because it is being done too often. They frequently come to my office, obviously dissatisfied with the procedure and the fact that the physician would not fix it. I feel disheartened by these encounters because, after speaking with these patients, I sense that they establish a relationship with these artificial devices placed inside of them. Their minds are focused on a procedure, and the procedure supersedes their own power over weight management.

There is a certain body-mass index point that the patient must "achieve" in order to receive the surgery. Therefore, if one gains enough weight, one can have surgery, a perverse incentive if ever there was one. Many of my patients have expressed sadness when they do not qualify for bariatric surgeries, claiming it is not fair that other people just a few pounds heavier are getting the quick fix. As we would expect, many people will not reduce their food consumption, alter their favorite foods or exercise if they know they can gain just enough weight to qualify for the surgery to fix their lives in an instant. Gaining a little weight is acceptable, even twenty or forty pounds, if they can ultimately lose the new weight and then some through a quick fix. Some readers may think this is a far-fetched scenario, but unfortunately it is well described in news reports, and I have had many patient interactions wherein this was the thinking. Some people appear to focus on becoming obese enough to qualify for the bariatric surgery rather than adopting a healthy lifestyle. These patients literally aspire to have the surgery rather than aspire to be healthy.

Although this is a delicate subject, my intent is one of sincerity, not disrespect. I feel compelled to raise awareness about this subject because these procedure are becoming more prevalent — even children are at risk of being subjected to them. Incredibly, bariatric procedures are being performed on adolescent children, and overall bariatric surgery rates have increased 800% in the last decade. It is now the most common elective surgery in America. These procedures are being recommended as a treatment for diabetes (obesity causes diabetes), and as an early intervention rather than as a last resort. Bariatric surgeons are also recommending that the indicated weight for these procedures be lowered so that more people, thinner people, qualify so the procedures can be used to save the country money in light of the new national Health Care Act. Excess weight and the associated chronic diseases such as diabetes and heart disease are rampant in this country because of unhealthy lifestyles: too much food, poor choices of food, and lack of exercise. I know that the surgeons will point to data achieved by rigorously scientific means that suggests the efficacy of these muzzles, but is it the best thing to do, since the patients must still adopt a lifestyle that they did not have prior to the surgery, a healthier lifestyle?

Why not offer the obese patient a real muzzle, maybe at least explain to him truthfully that this is what the surgery, in effect, does. It applies an internal muzzle. One way to begin an honest and healing relationship between the physician and the patient, as well as between the patient and himself, would be to explain the proposed procedure so that the patient can actually understand the exact imbalance causing his condition. The shock that would occur to the patient as he is discussing a muzzle may enlighten him to the true nature of his malady, which is a dysfunctional relationship between his physical being and his mind.

A mind that lacks an understanding of reality is a corrupted mind, and attempting to bring the patient back to reality after the surgery is all the more difficult. Convincing the patient to believe in the power of her actions when her anatomy has just been altered by major surgery or by placing a rubber band around her stomach is tantamount to a miracle perhaps. The focus of any discussion with the patient becomes the surgery, the complications of the surgery, and the actual versus promised effects of the surgery.

I find that obese patients are often dishonest with themselves about how they treat themselves and what they actually do for their own health. Usually patients' estimates with regard to the quantities and types of food they consume and the exercise they perform are inaccurate compared to objective data collected in a self-care log. This lack of a realistic view of how they are living will not change with surgery. Consequently, perhaps the worst of the outcomes associated with these bariatric procedures is that years will pass with missed opportunities to build a better relationship with themselves. In fact, these patients' poor relationship to food is not only *not* addressed through bariatric surgery but that relationship is also most likely going to worsen after the procedure, either out of a fatalistic acceptance of the failure of personal interventions prior to surgery or the validation of the powers of modern medicine if the procedure results in lost weight.

To make matters infinitely worse, the people who undergo these procedures will often suffer other ill effects from the same manufactured foods that caused their obesity. For example, consumption of manufactured foods causes inflammation, even in the absence of obesity, leading to arthritis, senility, loss of physical function, high cholesterol, high blood pressure, and cancers. So the fact that people may be able to eat poisons (manufactured foods) without the threat of obesity does not take away the fact that they are still poisoning themselves. In fact, after the procedure, if weight is less an issue, they are more likely to continue eating bad foods rather than following their doctors' orders about post-operative lifestyle changes. It stands to reason that bariatric procedures are likely to detract from the patients' perceptions of their role in their overall health because their focus becomes what the medical establishment can do for them rather than what they can do for themselves.

When we choose an internal device or medication as a method to health, we have not only become dependent on it but have developed a relationship with the artificial world in place of a natural means to our health and our life. I have chosen bariatric surgery as an example, but many more procedures would represent the situation as well. We are made ready for acceptance of this mindset by our upbringing within the Industrial Order, but our transmutation into pawns and inputs for the industry of medicine is complete when we accept their definition of health as being the "real" concept of health.

What the medical industry is really selling us in this new definition of health is the acceptance of chronic illness. This concept of health has vast potential to sell gadgets and manufactured interventions, which this definition of health tells us we must consume just like we consume the cereals in boxes, the factory-made donuts, the super-sized fries and Happy Meals. By this proposition, health becomes a modern day convenience delivered to us with many promises, but our instincts for health are lost. We tend to forget that health is not a scientific parameter or a static state but a dynamic continuum resulting from the resiliency and vital actions of an organism, a human being. We are sold something very different, and the transaction begins with the words that medical professionals use to entice us.

INDUSTRIAL RELATIONSHIPS

The prospect of manmade health is very enticing to our logical minds, especially when we feel ill, for this is but one more instance, it seems to us, of our technology making our lives better. Therefore, it is easy to see how we would be inclined to have an intimate relationship with an industry that is called, after all, "healthcare." But the industry's interest in creating relationships with patients (customers) before they are ill smacks of an attempt to increase market share out of thin air.

Doubtless, medical conditions will continue to be created or defined in such a way as to expand markets. Therefore, consumers (patients) need a new and special set of skills to effectively engage in their matters of health. Many aspects of modern medicine are not required for a healthy life, and indeed engaging with the medical industry has substantial risk. In the Appendix, I provide other helpful references to assist the reader in understanding the marketing creativity in the guise of healthcare, the slick advertising in combination with the illusion of scientific certainty that amounts to corruption born of greed that presents such an increased risk to the patient.

Unfortunately, for example, many endocrinologists are now recommending that virtually all people, even those with normal cholesterol readings, be placed on statin agents such as Lipitor. According to these healthcare providers, there is no level that is too low for cholesterol, which may sound reassuring and hopeful to some people, but the obvious question is: "Are we all born Lipitor deficient?" Apparently, we are all becoming Avandia deficient, Prozac deficient, and so on. The industry educates us that we lack pharmaceuticals and that is why we feel sick, why we have symptoms and diseases.

My patients are in fact subjected to potential harm when the medical industry wants them to feel comfortable, reassured that their cholesterol numbers are better, and yet these patients remain engaged in lifestyles that defy healthy living. When a powerful, licensed, expert "health" professional provides these types of remedies as a means to improved health, it is automatically implied that there is no other valuable way to achieve health or remedy the situation. The methods to address problems that are out of the professional's control are, implicitly, not worthy of consideration. This is intellectual dishonesty, but of course doctors are indoctrinated into the medical-industrial project from the beginning of their training, and so perhaps this amounts to bias on their part. Healthcare providers are taught that the answers for patients must lie in the evidence generated through the scientific method, the medical institutes and other proven means of science, and by virtue of the industrial identity most patients have formulated since childhood, patients come to doctors with expectations that industrial medicine will provide the best recourse for their ailments. The confluence of these two elements, patients' technology-entitlement mentality and the fact that healthcare providers have bought into their supreme efficacy when they use artificial interventions, means office visits are indeed loaded with over-diagnosis and under-utilization of common sense.

I can sense readers wondering about the benefits of especially aggressive and invasive procedures, those sold to us as lifesavers, such as cardiac catheters, which certainly appear to rescue many people. Many long-term relationships are built through those catheters, and of course there are benefits to them, most of the time. But with every positive relationship there often comes a negative one as well. The power of a rescue will assuredly influence the thinking of not only the patient but others who witness the event. The rescues alter the perceptions of the whole society, and we tend to all believe there is only good to be had from invasive cardiac procedures. Healthcare delivered in the rest of the developed world provides ample evidence to the contrary, however. Canada and the European countries perform far fewer cardiac procedures, and they have better health data than does our country, where, time and again, it has been shown that people undergo too many medical procedures.

The success of procedural and pharmaceutical rescues provides excellent opportunity to entice you to enter the world of modern medicine even before you feel ill. Advertisements and the internet expose us to several new potential diagnoses each day, usually presented with enticingly vague symptoms such as fatigue, abdominal pains, insomnia, or simply not enjoying life as you once did. Even our individual personality quirks are susceptible to being classified as medical conditions, therefore requiring drugs and special interventions to rectify. Our children are pharmaceutically deficient and cannot pay attention to us or teachers at school. If we sleep too little or too much, we are deficient in a drug. The symptoms being treated are multiplying as fast as conditions are being invented. Peter Conrad describes the medical industry precisely as one that transforms routine human conditions into treatable disorders in his book *The Medicalization of Society.*

Clever methods are also employed by hospitals to capture the hope and therefore affiliation (and eventually loyalty) from patients when they are well, so that when they are sick there is no debate as to where they will seek medical care. The potential side-effect is that people become cozy with an institution and an industry that derives its revenue from sickness, and therefore the likelihood of developing a symptom in need of medical attention becomes greater. In short, as we buddy up to the hospital centers when we are not ill, the threshold for entry into the health production process (hospital) has just been lowered.

I see this in community health screenings and seminars, where hospitals have often recruited specialists and aggressive physicians wielding new procedures to be guest speakers. One day I ran across a banner that stated, "Do something for your heart this month," and when I thought about the purpose relative to the content, I realized how diabolically creative such a simple statement is. We can indeed do much for our heart health, but attendance at these seminars has never been shown to improve one's health. And yet they are put on for "education" purposes by the same institutions that pride themselves on practicing only evidence-based medicine, institutions with business models premised upon illness rather than health. People will leave the seminars in the same health state they entered, but with one very important change: they have entered the health center's domain and therefore broken the ice. They have aligned their ideas for health to the ideals and methods of health promoted by the medical center. People are coaxed to enter medical centers for "health," and as I will show in subsequent sections, symptom evaluations (such as chest pain) have become a very lucrative part of their business model. Symptoms are more often than not just feelings and are far less likely to represent immediate harm or sickness at all, but this ruse has people entering medical centers more and more without actually being sick.

The "Do something for your heart health" seminar I saw advertised by a prestigious local hospital had cardiologists as speakers, professionals one would not characterize as motivational experts for healthy living techniques, but these people are quite motivational in their recommendations for heart catheters and other heart procedures. They sell procedures because that is how they earn a living, and yet they are recruited by hospitals to give lectures on "how to have a healthy heart." This is cynical at best. Naturally, potential patients are warmed up to the cardiologist's power over the heart and the proffered procedures become inevitable for the average participant, the threshold for the heart procedure thus lowered long before patients get their first twinge of chest pain. This seems predatory in nature.

Heart disease numbers will continue to grow in this country, but the number of heart procedures will in fact grow much faster than the disease it is supposed to be treating. The procedure rates can in fact grow orders of magnitude faster than disease rates because each patient can undergo multiple procedures before he or she actually succumbs to heart disease, the disease process that was never actually remedied in the first place. The ultimate ramification of industrial medicine on our heart health is, ironically, more heart disease because the reasons for the heart conditions are not addressed by modern medicine: the foods we eat and the exercise habits we need but desperately resist. Cardiologists have no vested interest in preventing heart disease because they are paid to treat it. The fact that performing more procedures on the population does nothing to abate the epidemic of heart disease among said population, especially in younger people, is important scientific evidence that is obviously ignored by patients, by the consumers of these procedures.

Several years ago, the COURAGE (Clinical Outcomes Utilizing Revascularization and Aggressive Drug Evaluation) trial showed that aggressive interventions, like stents, in patients with stable coronary artery disease did not improve morbidity and mortality outcomes for the trial's large study population. Patients are undergoing procedures that carry risks such as blood stream infections, ruptured blood vessels in the leg or heart, strokes, and heart attacks, and receiving little benefit from them. At the time the data from the COURAGE trial was published, my internal medicine partner and I predicted that the COURAGE trial evidence would be ignored by cardiologists in general. A recent follow-up study to the COURAGE trial confirmed the original conclusions, confirmed that cardiac procedures and even expensive nuclear stress tests are utilized to evaluate chest pains far more than necessary to make a meaningful difference in people's lives. Just as I suspected, however, both the original findings of the COURAGE trial and the subsequent findings have largely been ignored by cardiologists. The notion of an evidenced-based medicine practice has a peculiar meaning when some evidence can simply be ignored.

Procedures like cardiac angiograms can be helpful, but they certainly are performed with risk, and when performed unnecessarily, that risk is without justification. These procedures have become extremely common, nevertheless, and most consumers will not ever know if they actually needed one in the first place. They are likely to recall only the reports of benefits. I see this all of the time when, after a negative cardiac catheterization, the patient is found to have normal coronary arteries, he or she feels very reassured and not a bit unnerved at having just undergone a procedure that was not necessary. The reassurance these patients receive somehow discounts the negative aspects of overuse and allows them to continue as they were, eating the wrong foods and performing as little exercise as possible. Their risk of heart disease and the other modern maladies will continue to be high, despite reassuring test results. Indeed, a patient's relationship with natural health and his or her body does not usually change after a negative *or* a positive coronary angiogram. Sadly, it is generally harder to get anyone to engage in healthy conversations when she or he has just been relieved of concern for disease. The release from disease after a normal study is more likely academic or intellectual in nature, not physical. The odds are, however, that if the patient continues to possess deficient health skills, the next procedure will likely be positive.

It is interesting that so many people can whip out their wallet-sized photos from recent heart angiograms, their wallet-sized cardiac stent or pacemaker serial numbers and battery expiration dates, but they have no idea how many calories they ate for breakfast or the last time they exercised. For these patients, cardiac procedures equal a healthy life but their breakfast, lunch, snacks, and dinner are apparently irrelevant.

Using technology to peer into the dark crevices of our bodies that we cannot see ourselves is an excellent selling point, even if such illumination does not yield improved health. Our bias toward technology and our Industrial Order identity yield a nearly silly rhetorical question: how could it not be beneficial?

UNRAVELING OUR WORDS

Language, the defining trait of what it is to be human, is the most important tool that mankind has ever developed, but we use it with such laxity that it is difficult to realize its true importance. Products deemed "organic," for example, are not really created by natural processes but made in factories utilizing agricultural inputs that contain less pesticide residues or other quality specified by government agencies. Likewise, "all natural" is used to describe food but does not actually guarantee that the item purchased is made by nature. As I pointed out earlier, for example, maple syrup is generally not really made of maple, and the list would be exceedingly long if we considered all of the descriptive inaccuracies we are presented with each day.

Sadly, Americans have become accustomed to their words being hijacked by industries and bent or manipulated in significant ways to advance an agenda, promote products, and even incite large-scale cultural movements known as consumer trends. As I pointed out in "Acquisition of an Industrial Identity," many of our cultural values originate from and are fostered through corporate activities. We believe our industrial might is important, and therefore Americans tolerate widespread word abuse by industry. Indeed, we are trained as consumers to expect word bending and semantic play to benefit our favorite industries, conditioning that begins early, even as we are developing our language skills.

I am fascinated by this tendency from an anthropological perspective, but being a physician as well as a person concerned about his own health, my main concern is that such semantic misdeeds and language inaccuracies promote and sustain physical disease in America. This is not a trivial matter to a population that is now more prone to chronic disease than being prone to wellness. Indeed, it is possible that the quality of our lives as regards our physical well-being rests on our most ancient tool, our words, through a conscious re-examination of how we are sold our concept of health at the least. Hopefully, at some point, en masse we will demand greater honesty on the part of these advertisers.

Concerns over dishonest word-smithing are not particularly new. Politicians have been accused of serious word mincing throughout history. John Adams, one of our most influential early American leaders, was concerned that imprecise word use was a potential threat to our culture, our democracy, and to national cohesiveness. He was quoted as saying, "Abuse of words has been the great instrument of sophistry and chicanery, of party, faction and division in society." Indeed, industries and politicians both routinely utilize the synergism of psychology and media technology to influence our decisions on products and candidates, to direct cultural biases and behaviors, by bending meaning to their own ends.

Our current knowledge of psychological mechanisms relative to language, when combined with our technological prowess, specifically with message dissemination through mass media, provides industry the opportunity to warp meaning to serve its own needs on a mass scale. Because of continuous exposure to messages that include overtly programmed psychological seduction, our minds are under constant assault. Americans typically spend several hours a day exposed to messages from electronic sources, engrossed in their virtual worlds, and so it is no wonder that most of us are under the mistaken impression that the industrial food and medicine producers want us to have. This attention to these messages has a negative effect on our health in a more direct way too. We spend hours engrossed in various forms of media and therefore are less prone to engage in a more physical activities.

Self-serving messages may not do much harm when selling cars, boats, toys, and computers, but when words are hijacked by the food and medical industries to misconstrue vital aspects of our lives, the levels of chronic disease and suffering we see in our population should be expected. I propose that primordial matters, such as the health of a human being, deserve words that are chosen with care, concern, and precision. We should expect the medical industry to use very precise language to describe their interventions, precision in the service of accuracy and not misrepresentation, as is now frequently the case. Indeed, both the way the medical professionals, including the minions of the medical industry, and our government officials talk about healthcare should conform to a simple axiom: keep it simple and honest.

It seems laughable that the smartest creatures gracing the planet should require a great deal of intellect to discern what health actually is when our innate state of being and how to maintain it have been facts of nature since the beginning of human time. Indeed our convoluted notion of health has led us to a crisis, which one could argue is also a tipping point in favor of real human health. As noted previously, our children and young adults are becoming increasingly disabled, developing chronic diseases at increasingly younger ages, including diabetes type 2, high blood pressure, high cholesterol, and heart disease. Fully one-half of the population is engaged in daily consumption of a prescription pharmaceutical and twenty percent are taking three or more medications in order to maintain their vital signs in acceptable ranges. More than one in ten children is placed on daily medications, and this trend is worsening.

Young Americans are now disabled in such numbers that the military is having difficulty recruiting enough people healthy enough to serve in the armed forces. One out of three military applicants cannot pass the physical exam and fitness assessments for service, which means even our national security is being affected by the loss of youthful health. We are also facing potential national bankruptcy in part due to the costs of our medical care. The Congressional Budget Office predicts that, within the next twelve years, the equivalent of all federal government revenues will be consumed by Medicare and Medicaid, which amounts to less than half of the money predicted to be spent, the remainder coming from businesses and individual citizens.

This litany of bad news resulting from our current misperception of health could go on for quite a few pages, in fact. The point is that we shouldn't be satisfied with the state of "health" in America, and we should not wait for change within the current system. It is quite obvious that the word *health*, as used in the so-called healthcare industry, poses an imminent threat to the nation as well as individual citizens. We can demand the system change, demand the purveyors of lies cease and desist, but in the final analysis, we are going to have to recognize these lies for what they are and opt for a definition of real health in spite of the medical-industrial complex. I suggest the following as a starting point (and admittedly, this is somewhat reactionary in nature, a response to both the medical-industrial establishment's implicit definition of health and how our politicians talk about it): ***Health means people are well, more fit and therefore more productive. Health cannot put a country in financial ruin or place risk on national security. Health can only improve our situation, both as individuals and as a nation.***

In "Cellular Corruption," I discussed the integrity of our biological fabric as important to our health, but the integrity of our concepts of health ultimately affects our bodies and our overall health, too. We have seen how our concept of food has changed, now even including items of virtually no nutritional value and delivering us only harm. Likewise, our concept of health has been hijacked through industrial spin and obfuscation.

Indeed, if we place our hope for health on an industry that specializes in diseases, we will remain ill. We can begin to understand our collective misconceptions in this regard by evaluating the actual role of a licensed medical provider. As a physician and healthcare provider, for example, I am required to demonstrate expertise in the diagnosis and management of diseases, not health. I am licensed to treat diseases, not health. Physicians are routinely evaluated by state, federal, and other regulatory bodies in their ability to find disease in their patients, not health. As a doctor of medicine, I am legally liable for missing the presence of disease in my patients, but there is no liability for being remiss in an opportunity to cultivate health in a person. In short, I spend my days tending to the sick, fully immersed in increasingly complex systems to improve sickness outcomes, not providing health or instructing people on healthy lifestyles or otherwise serving health in any respect except the treatment of disease — at best a very limited definition of the concept of health and at worst an excuse for the medical-industrial complex to dissimulate.

Even Hippocrates, the father of medicine, did not study and attend to healthy people; his was interested in and paid attention to sick people. In the medical school curricula around the nation, we have courses such as the Mechanisms of Disease, but there is no course entitled the Mechanisms of Health. I suppose the medical industry is called the healthcare industry because it would sound terrible to call it the sick-care industry, but this is in truth what a physician spends the day doing, diagnosing sickness and treating the sick.

One day I randomly selected eight people on a local street for an unscientific study. I told them I was interested in their concept of health and healthy living, and I asked them, "Over this next year, what would you do to achieve a better state of health?" The answers were similar for all respondents: eat better foods, sleep better, exercise, manage stress, and become more spiritual. None of them said they would go to the doctor more often or check themselves into a hospital as a means to stay healthy. It is interesting that uniformly absent from their opinion of how to secure health was enhanced interactions with healthcare facilities. There is a fundamental difference between the methods people believe are necessary to achieve better health and the methods to health that are sold to us by the most prestigious evidence-based medical centers and pharmaceutical companies. It is also interesting to note that it is not necessary to have lots of medical training to discern the true path to health, but it does require recognizing a bill of goods when one is being sold, a manner of technological intervention devoid of a focus on diet and exercise.

People can plainly see that health arises from nature and is the default state of the vast majority of life, but they deserve to know that health must be cultivated from within us with the same scientific certainty that a physician diagnosis disease. A rescue from sickness, provided by the medical community, subsequently requires the reinstatement of healthy processes through natural means. That is, patients must establish healthy behaviors after a medical emergency is averted. Cardiac by-pass surgeries and other aggressive medical interventions still require the patients to cultivate their health independent of the surgeon's skills. If they do not, the surgical skills were unnecessary as the patient is simply creating the conditions for the next intervention and is disabled in the interim. Patients with heart disease still have heart disease after heart surgery, but they can still pursue health as a goal independent of the procedures and heart condition.

It is easy for those suffering from a chronic disease to become disillusioned. They interact with the medical industry a sizable percentage of their time, seeking a healthy state, but they get a chronic condition instead. Diabetic patients are coaxed into checking their blood sugars four or more times a day with the hope of better health. They will then adjust insulin injections and their medications according to the blood sugar readings. The reason posed for this intervention is that better sugar control will provide an enhanced functional state of life. Diabetics tend to believe that, if their blood sugar readings are normalized, they might live as if they were *not* diabetic. Unfortunately, this is not the case. Patients remain diabetic and still have serious risk for all of the major associated diseases, including Alzheimer's disease and cancer, despite control of their blood sugar.

In fact, scientists have found that tight control can harm more people than it helps because low blood sugar puts people at risk of coma, seizures, heart attacks, and strokes. In fact, the ACCORD (Action to Control Cardiovascular Risk in Diabetes Study Group) trial, looking into the benefits of intensive therapy, was stopped early due to excessive deaths. Furthermore, tight control of diabetes has never been proven to lower a person's risk of a heart attack, stroke or Alzheimer's disease, those being perhaps the most concerning aspects of diabetes. If diabetic patients survive long enough with tight glucose control, they may lessen the chance of kidney failure (and therefore dialysis), blindness, and neuropathy (painful feet), but the odds of having such problems are still much higher than for non-diabetics. The products sold to diabetic patients, presumed better health by the aggressive treatment to normalize the blood sugars via the latest treatment and technologies (glucose meters, testing strips, medications, and doctor visits), is not real health. Instead they receive a chronic disease management plan, which is suboptimal for the patient. By the time a patient is diagnosed with diabetes, the cellular corruption leading to the disease has already begun to corrode the body, and the ability of science to ameliorate the tissue destruction is quite limited, much more limited than what doctors impart to their patients. The most optimal treatment, and the best hope for a healthy life, lies in the power of the patients' actions rather than aligning their lives to medical centers.

The best outcome is not to be diabetic in the first place. This simple truism is not forcefully espoused by the medical establishment, however, at least not as forcefully as it advocates for artificial interventions after the fact, despite more than one large study showing what I have always believed to be the case: the type 2 diabetic condition is at least 95% preventable through a healthy diet and exercise. Echoing this sentiment is a recent report by the National Institutes of Health, which states that diabetes type 2 is 80% preventable, even within the context of having a significant family history, a low estimate because many study participants do not actually undertake the necessary lifestyle changes. Even after being diagnosed with type 2 diabetes, most people have very good options, such as lowering their medication burden (opting for a milder form, so to speak) or reversing diabetes completely, but much of the success of lifestyle change on the patients' lives is a function of their level of desire for better health.

In fact, in my professional experience diabetic patients benefit far more by going to the gym and focusing on their diet interventions than focusing on logging blood sugar checks. Many of my patients have been surprised when, during their mandatory quarterly meeting to review their diabetes control, I recommend against doing their blood work, the recommended standard of care, because they have lost sufficient weight and are taking part in an exercise routine. I explain to them that I do not wish to place the focus on what the medical system can do for them because they are capable of doing far more for themselves. I explain that the recommended tests by the American Endocrinology Society are superfluous and unnecessary when diabetic patients do what they must to change their lives.

Living a healthy life is in some sense a concept separate from our afflictions, tests, procedures, and even disfigurements. Our attention is best placed on the optimism inherent in living well. No matter how normal our test results may read on the laboratory printout, we will never be healthy by continuing the behaviors of self-harm that perpetuate cellular corruption within our bodies. Normal tests do not indicate a healthy life any more than abnormal tests indicate that one is absolutely sick. The certainty that people conjure in their minds as they mull their recent doctor appointments, tests, and medication refills is not the certainty that actually exists in their lives. There is a great deal of discrepancy that can only be reconciled by acting healthy, which is as simple as being active and eating truly natural calories.

In order to have a better, more honest relationship with their patients, medical professionals must use words that may lack appeal from a marketing point of view but which offer a more accurate description of the services provided by the medical industry. A good start would be to admit that physicians provide sick-care. People would then be free to pursue actual health, true health as opposed to artificial intervention for its own sake or the cascading health effects of said interventions, by natural means.

PREVENTING HEALTH

Lies and obfuscation never breed optimism, even if the words are meant to be optimistic. Preventive Health or Preventive Medicine is the more optimistic (sounding) branch of industrialized medicine. The choice of descriptor imparts a distinctly nurturing implication and a proactive feeling. Unfortunately, preventive health is actually another form of established disease management. Aside from immunizations, the procedures and methods generally undertaken are to detect and hence manage disease in earlier stages (disease burden reduction). True prevention of disease, at least according to my understanding of the English language, would mean that the disease is not present and therefore has actually been prevented, not managed postdiagnosis, no matter how early that diagnosis occurs. If a disease is not present, then we would not benefit from the so-called preventive health screening tests that look for the disease.

Indeed, when we follow the mandates of industrial-made health, when we subscribe to their definition of healthcare that is really sick-care, we are likely to be subjected to unnecessary tests, unnecessary procedures, and diagnoses that are misleading. Every test and procedure has intrinsic error rates, and so we must remember that there is a chance that disease is being detected that is not there, and, rather than being prevented, the disease is actually being treated when in fact the disease is absent from the patient. Before we embark on a review of the individual tests that are typically associated with preventive health practices, it is important to understand the context of our relationship with the healthcare industry that espouses only good will come from our being subjected to these procedures, or at least more good than harm.

There is in fact no evidence that modern medicine can make a healthy person healthier, despite the cozy words of preventive healthcare practices. It is the manner in which a person lives that either elevates the risk of chronic diseases and cancers or lowers that risk, not the lack of colonoscopies, mammograms, blood tests, or CT scans. To sustain health we must be clear about these things, otherwise we may sign-up for "preventive health" and actually find out we are instead preventing our health by being pulled into the medical industrial system, where one procedure leads to another, where many prescription medicines cause some side effect that is then in turn treated by another medicine — or worse, in that there is actual complicity from the healthcare providers with regard to unhealthy behaviors. Many healthcare practices — perhaps most, especially those labeled preventative — are performed out of fear, curiosity, or guilt but not necessarily out of scientific reasoning. For instance, many people spend their own money (as opposed to the insurance company's money, because they have not received a diagnosis that would require a test) and are being harmed by the performance of the total body scans, and they do so to allay fears of unknown disease.

The annual physical exam, while certainly less potentially harmful than a scan, is also a "preventive health" ritual without scientific backing. However, despite there being no evidence to support claims of efficacy for such exams, Medicare has been persuaded by physician groups to cover yearly physicals. As Allan Goroll, MD, Professor of Medicine at Harvard Medical School, explained in the American College of Physician's newsletter, "ACP Internist" (Jan 2010), there is absolutely no evidence that this expenditure of time and resources improves patients' lives.

Certainly, people seek medical screening tests to secure some degree of certainty over their health fate or because they were told to do so by our society (now celebrities tell us we should get tested in so-called public service spots). Placing our health in the hands of the medical industry when we are gravely ill is a situation in which the gains are likely great and the risk of loss is more acceptable. But a much higher level of proprietary scrutiny is in order when we are looking to maintain or improve the health we have. If we already have excellent health or health potential, we have substantial health to lose, therefore we should require more clarity and honesty in our interactions with medical professionals because modern medicine is capable of diminishing the health that we already have. I find that the most detrimental action by the medical industry is the persuasion to use artificial interventions that divert our attention from our role in our illnesses and our role in personal health. This serious problem is then compounded by the threat of tests causing us harm, over-diagnosis or worry.

Do healthy people benefit from any screening tests? The answer to that question is obvious when one realizes that there are no screening tests proven to reduce disease in people with excellent health habits. If someone has extremely healthy behaviors, then the likelihood of sustaining serious side-effects attributed to virtually any screening test becomes an unacceptable risk because the benefit of finding disease that is present but not known is too low to undertake the risks of these procedures (I will explain momentarily). Because all of the large studies performed to assess whether or not a procedure or medical intervention will improve someone's chance of surviving disease have been performed on "average" American people, who are by definition unhealthy people, we should not be generalizing this data to a healthy person. That is, any random sample of Americans enrolled for study purposes will represent the statistics of the overall population of the nation:

sixty-seven percent will be overweight, and of the remaining third many will most likely have unhealthy lifestyles (smoking, excessive drinking, diets without fruits and vegetables, or diseases not associated with obesity). By definition, healthy people should not be overweight or suffering from chronic diseases (even if they are treated for them), should follow a diet rich in fruits and vegetables and low in manufactured foods, and perform regular vigorous exercise. There are so few extremely healthy people these days that there is no way to study them with statistical confidence (one needs large numbers of people) to recommend tests with any scientific certainty. Studies performed on unhealthy people, no matter how well they are performed and no matter how sound the statistics are, yield results that cannot be applied to a different population, a healthy population whose lifestyles no doubt indicate that healthiness in the first place. Despite the auspices of preventive healthcare, we find that people must already be sick enough, through their lifestyle, to benefit from the screening/preventive health tests. Then to repeat the tests in the future, they must remain sick. That is, if there is any benefit at all to be gained. Furthermore, there is no incentive to study truly healthy people because there would not be enough disease discovered to justify the cost let alone to recoup the money spent for the majority of screening tests. Likewise, however, people with multiple medical problems or a poor prognosis (sicker people) need not be tested because longevity or one's overall health can't be sufficiently improved to warrant the risk of the tests in question. In summary, to benefit from preventive health methods, one must be substantially sick, perhaps an average person similar to those people enrolled in the trials, but not so sick that the tests are too risky to be done in the first place. Where in this scenario is the concept of "preventive" medicine or any optimism as to one's future health?

In fact, people are routinely having their health stripped from them by screening tests because of the complications associated with undergoing subsequent procedures. These patients have outcomes such as radiation exposure and perforated bowels, but the most dreadful is the diagnosis of disease, especially cancers, that are not present at all. Over-diagnosis of disease is in fact disease creation, or health prevention, in the name of preventative healthcare.

Over-diagnosis of cancers is a common problem in the new push for more screening tests. In a recent article in the *Journal of the National Cancer Institute*, Doctors Gilbert Welch and William Black estimate that, while screening one thousand women for breast cancer over the course of ten years, up to ten women will be diagnosed with breast cancer, and subsequently treated as if they have breast cancer, to save one person from death from breast cancer. Furthermore, between five and fifteen women will be diagnosed with an early form of breast cancer, but it will not impact the course of their lives, and then another set of women, between 200-500, will be subjected to needle biopsies to allay fears of a breast nodule. If we consider the trend in death rates from advanced stages of breast cancer over the last three decades, we find more uncertainty with regard to the widespread recommendations for breast cancer screening by mammography. In the November 21, 2012 edition of *The New England Journal of Medicine*, Archie Black, MD, and Gilbert Welch, MD, conclude that "[a]lthough it is not certain which women have been affected, the imbalance suggests that there is substantial over-diagnosis, accounting for nearly a third of all newly diagnosed breast cancers, and that screening is having, at best, only a small effect on the rate of death from breast cancer." Obviously, the power of mammography as a means to save lives is not as strong as advertised.

This bit of uncertainty, the negative aspects of testing, does not garner discussion in the rooms of a doctor's office. A prescription for a test or referral for a procedure is much easier and satisfies the fears from both sides, under-diagnosis. Over-diagnosis carries much less risk of litigation for a physician, and the patient will usually never know of the problem. A simple question that can get to the heart of our screening debates is: "Will the screening test in question improve or detract from my ability to secure a healthy life?"

Over-diagnosis of disease is a terribly unhealthy situation for our entire society and likely to become worse with more technology entering medicine by the day, a problem unique to America, where medical tests are ubiquitous and well-funded. If we over-diagnose substantially, then the result is more fear within society, and we are prompted to give up more money and resources to fund more screening tests and more therapies. The success rate for screening tests and then cancer therapy to spare someone's life from a cancer that he or she does not have will always be excellent for these patients who never had it in the first place. I suspect that the medical community would justify any over-diagnosis in the attempt to save that one-in-a-thousand life, but there are substantial articles and books on the subject of over-diagnosis and over-treatment occurring in medical centers across America, and almost no location is unscathed by this dilemma. This is a new form of plague, another uniquely manmade plague, one that arises as we try to tame our biological risks with laser precision.

Common tests and procedures associated with preventive medicine include mammograms, prostate specific antigen assays (PSA), cholesterol tests, blood pressure readings, sugar measurements for diabetes, pap smears, colonoscopies, and cardiac stress tests. Aside from immunizations, there is actually very little in the way of disease prevention from these so-called "preventive health" interventions.

Mammograms do not prevent breast cancer but can identify cancer that is already present, if they work as intended and do not over-diagnose women with cancer. Many people value mammograms but miss the message that the true prevention of breast cancer is by maintaining a healthy weight, eating a low-fat diet, and exercising regularly. Prostate tests, either the PSA or digital rectal exam, do not prevent prostate cancer, but eating a low-fat diet and ingesting lycopene, which is found in tomatoes, regularly does prevent prostate cancer. The benefits of routine screening by PSA tests have been refuted, in fact, and screening tests are not recommended for cancer detection. The reason this change in policy came about is that testing for elevated PSA levels in healthy asymptomatic men is more likely to cause harm from procedures to confirm whether or not cancer is actually present than living with the cancer. It is interesting to note that Britain was ahead of the US, the self-professed leader in healthcare, in spite of evidence to the contrary, in recognizing that prostate cancer screening was hurting more people than it helped. Instead of developing a relationship with PSA tests and urologists, a man can develop a better relationship with himself and improve his odds with regard to prostate cancer.

Performing cholesterol tests and taking blood pressure readings and sugar measurements have never been shown to prevent high blood pressure, high cholesterol, diabetes, stroke, or heart attacks. In fact, as pointed out previously, despite increased vigilance and increased specialist care over the last few decades, the number of people being affected by these diseases, and especially the young people in this country, has increased to epidemic proportions. Testing for these conditions is therefore of little consolation to the next generation of young people that will suffer from diabetes, kidney failure, disability, heart attacks, and strokes. True prevention of these diseases will not be achieved by more tests, newer tests, or more thorough applications of these tests. Optimal results will require a healthy weight, regular exercise, a low saturated-fat diet, routine ingestion of antioxidants from fruits and vegetables, adequate omega-3 fatty acids, and a positive outlook achieved via a less stressful life.

Pap smears do lessen the conversion rate to cancer when the cervix is infected by the Human Papilloma Virus, but these tests have historically been over-utilized. The way to prevent cervical cancer is to vaccinate against the Human Papilloma Virus, use condoms, and/or practice monogamy. Sexual promiscuity increases the risk of cervical cancer as well as other infections like HIV, syphilis, genital warts, and gonorrhea.

Colonoscopies do not prevent colon polyps, the earlier stages of cellular corruption before complete cancerous conversion, and questions still remain today about whether this test should be performed as a mass screening measure. Colonoscopies are recommended to many healthy people who do not benefit from them, and studies are now confirming this idea. A recent study published in the October 10, 2012, edition of *The American Journal of Medicine* indicated that women under seventy years of age should be screened by other means. Furthermore, the June 21, 2012, edition of *The New England Journal of Medicine* contained an editorial by John M. Inadomi, MD, that also alluded to serious concerns inherent in the use of colonoscopies for widespread screening of healthy people. No randomized, controlled trials have yet demonstrated that colonoscopies reduce cancer mortality, and for this reason the United States Preventive Services Task Force (USPTF) has yet to endorse colonoscopy as the best screening test for colon cancer. The

USPTF maintains that other screening measures (which are still looking for disease that is already present) such as the smaller colon scope (sigmoidoscopy) and the fecal occult blood test are excellent screening tests with less risk of colon perforation, fewer side effects (including a day of missed work), and lower cost. The risk-to-benefit ratio appears to be different from what the gastroenterologists who perform colonoscopies tell people.

I routinely see people instructed to have follow-up colonoscopies for polyps or inflammation that are not likely to diminish the quality of their lives, and along those lines the National Cancer Institute recently determined that follow-up colonoscopies for colon polyp surveillance were being performed much too often. Within the paradigm of evidence-based medicine, not all evidence is used for the patient's benefit. Much of the scientific evidence that is inconvenient or less apt to generate revenue and sensationalism will usually be ignored. Truths are more often chosen to be true, not necessarily proven to be true.

Colon cancer is linked to the American diet. Bile acids, which are normally produced by the liver and secreted into the intestinal tract to aid digestion of fats, are therefore increased in response to diets containing more saturated fats. They are linked to colon cancer because they are caustic to the cells of the gastrointestinal lining and are thought to induce cellular changes leading to polyp formation and eventually cancer. Other factors, such as a reduction of physical fitness and alterations in our normal gut bacteria (flora), further enhance bile acid irritation. Widespread antibiotic use is also likely creating colon cancers because the medications kill the beneficial bacteria in our guts and therefore induce chronic inflammation of our colons and weaken our immunity against cancer in the intestines. Fiber, found in fresh fruits and vegetables, is thought to reduce our cancer risk because it binds to and pulls these bile acids from our digestive tract and also improves our gut flora. Antioxidants that are plentiful in fruits and vegetables bathe the colon cells with cancer-fighting compounds. Therefore, a diet rich in plants and lower saturated fats will lessen the risk of colon cancer by at least four independent mechanisms. Furthermore, we have the ability to lessen our colon cancer risk even further (a powerful and independent mechanism) by staying active. The National Cancer Institute states that exercise performed routinely will decrease one's risk of colon cancer by 50% without the need for tests or medical therapies. The recommendations for colonoscopies as cancer screening tests are therefore a less potent method of prevention compared to adopting a lifestyle that is not prone to colon cancer in the first place (true prevention). In effect, we choose our risk of colon cancer. All of the debate over whether or not and by which method to be screened for colon cancer distracts our minds from its primary purpose: to serve the body well, securing wellness first from within.

There are other maladies of the colon linked to the same lifestyle habits that promote colon cancer, such as diverticulosis that has become infected, causing perforated bowels similar to appendicitis, and ischemic colitis, which is bloody diarrhea and pain due to the lining of the intestines dying off because it lacks adequate blood flow (clogged arteries). Both of these conditions are more common than colon cancer, and colonoscopies cannot prevent them either. The same lifestyle habits that prevent colon cancer also prevent these dreadful colon problems. The bottom line is that having a relationship with gastroenterologists may not be the best way to live a life free of colon cancer and improve one's bowel health. We can do better than the experts will tell us.

There are no screening tests for ovarian cancer, pancreatic cancer, and lung cancer, but there are disease prevention measures like abstinence from smoking, maintaining a healthy weight, and eating a low-fat diet rich in fruits and vegetables. Pancreatic cancer is most associated with chronic pancreatitis (usually alcohol induced), smoking tobacco, obesity, and diabetes. Prevention is what the patient does for himself or herself, which is infinitely more than what medical centers can do.

As you might have surmised by now, our personal power over disease is so great that it not only prevents diseases but also prevents the need to succumb to the procedures, tests, and recommendations of the most powerful industry on the planet, the healthcare industry. A proactive and healthy life, a lifestyle that will make us vital people, wields far more power over our health than relying on tests to search for sickness that is yet to be found. We alone have a raw power that can only be fully appreciated if we know where our preventive health lies, and therefore the distractions offered by the medical industry, including so-called preventative procedures, are important to recognize lest they diminish our power to achieve real health.

LOST IN TRANSLATION

Chronic disease equals chronic profits for the medical-industrial complex. As our society accepts a blurred distinction between wellness and sickness, a distinction developed by the medical industry and propagated by it at every turn, a competition of ideals arises within the individual: natural health, which every individual seems to understand as a function of being active and eating right, and artificial health, which is the application of endless interventions. This competition of ideals results in, to use the words of Thoreau, the "quiet desperation" of modern man. We are, in other words, living a less fulfilling life by virtue of our enslavement to the Industrial Order, which promises health but delivers only chronic illness, and we know better. We know that the way to health is diet and exercise, and we know it in spite of the onslaught of messages we receive daily from the medical-industrial complex.

If a person's behaviors include eating excessive calories and remaining sedentary, then he or she is likely to be undergoing cellular corruption, which is slowly corroding the tissues and increasing the risk of cancer, diabetes, heart attacks, and strokes. These processes are occurring years before the evidence shows up in test results, and so I caution people who have good test results but who otherwise show signs of poor health and disease risk that just because their laboratory tests are in the normal range does not mean they are healthy or can avoid these afflictions. In short, I see unhealthy patients inadvertently reassured by their test results. Patients may come to see me because they know they are not living a healthy life or they simply don't feel well in general, but upon receipt of blood test results that are "normal," they then assume that their condition is not as bad as feared. If I know as their physician that some of my patients are physically unhealthy, how shall I motivate them to change behaviors that are hurting them with tests that indicate they are normal?

Studies show that the best way to predict anyone's future mortality is to look at his or her functional capacity. The higher the functional capacity physically, the lower the risk of all causes of disease, even mental illness. This form of testing is generally done by the person and will likely signal the earliest stages of organ dysfunction (or better still: generally prevent dysfunction). Testing our functional capacity ourselves is certainly a more sincere form of health evaluation than passively awaiting abnormal test results and relying on them to provide us with information about our personal status. Self-assessment of our functional status is obviously a better proposition. I suspect that, upon realizing that we are low-functioning physically, we are more likely to solve the dilemma by means that would enhance our physical capabilities and not resort to drug therapy. But, as we should expect within the Industrial Order, technology breeds additional technology. Testing by artificial, less functionally natural, methods greatly enhances our expectations of artificial interventions.

An emblematic example is Curt, who was my patient for years. He came into the office for a yearly visit like clockwork, always joking that he needed to see how I was doing once a year. In reality, his company (Curt worked for a local packaging company where he held a managerial position and received good corporate benefits) paid for the full physical exam and tests, and so, he told me, he "might as well take advantage of them." Occasionally, I would see him between these visits for a cold or something else minor, but overall, he did not need the services of a physician.

Every year I would invariably find that Curt was carrying a few more pounds around his waist, and he would always tell me that things would be different the next time I saw him. On one particular visit, he had obviously gained considerably more weight and appeared to be bothered by it. He also had borderline high blood pressure readings, and I sensed he was becoming more concerned about his health circumstances. Before we embarked on the usual gambit of tests, I offered him a diet regimen and recommended some exercises that would be easy to do and would make a difference to his health. I also offered him weekly office visits to reinforce his behavior and so he could remain engaged with me and therefore stick to his diet and exercise. I wanted him to do this for three months before we drew the blood for tests so he could increase his physical activities and be a more active participant in his health.

The tests were superfluous since I already knew, and I suspect he did as well, that his behaviors were adversely affecting his health and were in need of change regardless of the test results. As I noted in my own circumstances, we can obviously see an unhealthy state in the mirror, on the bathroom scale, and in the facts that we cannot run as well as we used to and do not feel as well as we once did. A loss of physical performance obviously indicates dysfunction, but what are also not generally associated with a substantial physical dysfunction are the nights of poor sleep, the increased stress reactions, and the generalized stiffness and aches in our joints.

Every year I would review Curt's blood tests with him, and they were never normal but not terrible either. He usually had mildly elevated readings across the board, and he would promise that next year's results would be better. Those better test results never materialized, and I finally asked Curt why he wanted to perform these tests every year if he did not intend to change his habits. The results of his blood tests seemed to have little impact on his unhealthy lifestyle, nor did they seem to instill any sense of motivation for change. The year I thought we should focus more on lifestyle changes before performing the tests, Curt was carrying an extra 35-plus pounds and did not appear to be a person who would enjoy jogging a couple of miles. Nevertheless, because he was a vibrant character, I had the impression that I could motivate him into changing his lifestyle.

Curt insisted we perform the tests before discussing things further. This round of tests, unfortunately, placed Curt into the metabolic syndrome classification. His systolic blood pressure readings were now consistently above 130 mm Hg, his HDL cholesterol below 25, his triglycerides above 150, and fasting blood sugars above 100. Curt's waist circumference, which was 38 inches, should have told him all he needed to know without these tests. Curt did not feel particularly different from how he had for the last five years, however, although he said he recognized that he had gained some weight.

According to the National Health and Nutrition Examination Survey (NHANES), as of 2006, 34% of US adults meet the criteria to be classified as having the metabolic syndrome. The American Diabetes Association claims that more than 64 million people in the US are classified as such. These patients have substantially increased risk for heart attacks, strokes, blood clots, kidney failure, and a fatty liver that may give them cirrhosis and liver failure. There is also a statistical correlation between this condition and degenerative arthritis and most common cancers. If allowed to continue down the same ill-health path, patients become diabetic and face multiple-organ dysfunction too.

I informed him of his new diagnosis, the metabolic syndrome, and provided him with this data. I hoped this would provide more urgency to Curt regarding his physical health concerns and risks in spite of all the years I had tried to get him to exercise and eat better. He asked me very sincerely, "What are my options?" I reiterated the lifestyle-change outline I had provided in the previous visits, and I told him that studies show that just a 12-pound weight loss can reduce one's risk of diabetes by more than half and a 20-pound weight loss reduces that risk by 80%. I pointed out to Curt that some of my patients had even postponed the diagnosis of diabetes for years by improving their lifestyle. Curt was playing golf every couple of weeks, fishing when he could, and almost never missed work, and therefore I knew he was capable of more routine physical activity. I tried to motivate him to engage in lifestyle interventions before considering medicines.

Looking back on this scenario, it seems as though Curt had been preparing for this day for quite some time because, as I was speaking about those interventions, he quickly pointed out that he had seen advertisements in magazines about medications that could treat the metabolic syndrome and that he had friends who were on medication for it. Some of the medicines that are used to treat the syndrome, Avandia for instance, can cause weight gain and swelling, but an older medication used for diabetes, metformin, can actually induce mild weight loss as the body is more efficient with insulin use. Curt told me he wanted a prescription for metformin because his friend had experienced good results with it. Since he met the criteria for the use of the medicine, following guidelines set forth by the American Endocrinology Society, I prescribed him the medication, but I made clear that he did not have to be on the medicines forever. If he shed ten pounds or so and became more active, he could potentially go off the medication. Four months later, Curt came for a follow-up visit and he had lost three pounds, but was not exercising and his eating habits had not changed substantially. Because his blood work had improved a bit, we decided to continue the medication for several more months and recheck the blood work at that time.

Curt had come to me for health, but he actually refused to take that path, embarking instead on the same path in life, but with medications in hand. He could have begun to be healthier at any point in this exchange, and he could have begun to be healthier immediately. As soon as one takes that walk around the block or shows up to the gym for the first time, real health is already improving because the person is engaged in being healthy. Taking medications and ignoring one's functional capacity is not real health but a numbers game that does not yield real results in terms of living better. The charade of the medical industry is to treat numbers and then to tell the patient that they are healthier because their numbers are better, never mind their level of functionality. Even when patients are given the choice, they tend to choose the Industrial Order over a more natural, and more powerful, means to health. Curt certainly had a choice, and he quite consciously chose to give his life over to the Industrial Order, transferring his power in matters of health to the medical industry and its artificial interventions. Curt was lost in translation, his biological attributes first altered by industrial foods, and then later willingly transferring his health to industrial medicine.

If we follow the Industrial Order and it creates disease, only to continue with the Industrial Order to rescue us from disease, then what we have really preserved is the Industrial Order — at the expense of our natural endowment of health.

Patients frequently confuse preservation of their Industrial Order identity with self-preservation, which is understandable, given the power of the industrial identity and the power of the forces reinforcing it. My point is that our health belongs to us and not the healthcare system. Health should be viewed as the most important form of self-expression because we can do more for ourselves than the industry of medicine can do for us. Our health is our personal, biological identity, and if our intent is to have good health, we must diminish the influence of the American culture on our minds and bodies, lest we become people we do not wish to be.

We understand intuitively that self-preservation actually means the preservation of our original selves, our true selves, our natural selves; but few of us manage to bring that intuition to the level of conscious action. Once that step is taken, however, the results are extraordinary. Because I now appreciate the biological nature of my life, I know to my very core that self-preservation is the preservation of my natural attributes, those natural tissues and systems that I was born with. I choose to belong to the Natural Order more than I belong to the Industrial Order, and if I can do it, so can others. I will share what I mean.

THE
NATURAL
ORDER

THE ETHIC OF SELF-PRESERVATION

Tao gives life to all beings.
Nature nourishes them.
Fellow creatures shape them.
Circumstances complete them.

— The *Tao Te Ching* of Lao Tzu, number 51,
translated by Brian Browne Walker, 1995

The physical laws of nature that govern the fitness of an organism, an ecosystem, and the entire biomass constitute the Natural Order, and therefore it follows that the Natural Order, at one level what Lao Tzu called the Tao, creates all life. Over the course of approximately four billion years of uninterrupted life on this planet, the Natural Order has proven its ability to sustain living organisms, and indeed, the fates of all organisms are intertwined. Since life arose on planet Earth, the entire biomass that has ever lived has subsisted on products assembled by other organisms — and not mankind's innovative foods. Another mechanism by which the Natural Order guarantees that life will go on is physical struggle, which is the source of an individual organism's strength and a species' adaptation for fitness over generations.

As living organisms, we are responsible for our fitness both as individual organisms and as members of the human race, and because we are products of the Natural Order, we have an obligation to follow the dictates of that order. And we can summarize the two most important components of the Natural Order for human beings this way: eating natural whole foods (those produced with minimal human intervention let alone industrial production techniques) and engaging in strenuous physical activity.

Surely, we can marvel at the Natural Order of life and the complexity of our biological processes, but to sustain a healthy life we must also respect these processes and live within the design that is the Natural Order. It seems miraculous that such a simple plan works so well. It is also miraculous that we are surrounded by so much proof that this simple plan works and yet persist in living sedentary lives consuming manmade foods. Less complex animals than ourselves seem to enjoy healthy states and a certain biological authenticity that many of us dare to dream of; and indeed it seems obvious that the majority of non-human living organisms on Earth have a higher percentage of thriving individuals with optimal health than do the humans living in American society. Furthermore, the vast majority of less complex animals also enjoy optimal fitness and health without the interventions devised by mankind's intellect. It seems an easy conclusion that living in accordance with the Natural Order makes life vital. As has been argued thus far in this book, health is not an invention but a spontaneous arrangement of biochemical energies and materials, the word *health* a synonym for life itself, a distinction that I did not fully understand for the majority of my life.

I like to think of the Natural Order as providing the pathways (physical and even cognitive) necessary for my life, but because my life actually arises from these pathways, the Natural Order is also the context within which I live my life, the perspective from which I attempt to withstand the effects of the Industrial Order that would have me live an other-than-healthy life. Most people believe society is the primary context for their lives and, for most Americans, the most influential force; but that would only be possible if societies could be built without our biology. In fact, therein lies the crux of our healthcare debacle: modern humans live almost oblivious to the biological context of human life, not mindful, let alone respectful, of the Natural Order.

In truth, a human life is simply a compilation of internal molecular consequences bringing order to biochemical molecules creating larger structures such as cells. Still larger relationships develop between cells, tissues, organs, and then outward, between other beings, and ultimately with the entire universe. Humans function best when they are not separated from the Natural Order, not disconnected from their natural being by an Industrial Order identity.

A Transition In Mind

Until 2006, my life did not feel very engaged with the biological world. My behavior indicated that I was no more concerned about being connected in some fashion to the biological environment around me than I was with the biological environment within me. During that year, I began to hear more discussion about global warming and the planet's loss of biodiversity, but those issues seemed extremely foreign. Besides, there were so many other seemingly important matters to consider. When I realized that I weighed more than 200 pounds (and at only 5 feet 7 inches in height), I also realized that I had disconnected from the concept of survival of the fittest. I further realized that the society in which I live and my professional success had confused me about the importance of my own biology to my life and my connection to the natural world.

My body and mind were corrupted by the Industrial Order. I felt physically unfit, spiritually bankrupt, and mentally bored. The preservation of my natural gifts, both body and mind, was not a priority, and certainly not an intellectual interest. Only after I gained a new reverence for how human life should unfold could I focus on my spiritual, mental, and physical essence, addressing those individually but also collectively. After all, this is what being a human amounts to, this confluence of our inner and outer lives. It became clear to me that I had stepped about as far as one can outside the very context of life — the Natural Order.

Before my awakening to this all important fact, conversations overheard at Starbucks about concerns, including global warming and sustainability, were in the same class as hearing somebody talk about visiting northern California: the conversations of would-be hippies and tree-huggers. After a certain amount of shock to my ego, however, I realized that I am just another organism residing on planet Earth and my health is dependent upon my connection to the natural world.

Coincidently, my wife and I opened a medical spa in Bountiful, Utah, at about the same time as I was coming to my realizations about the Natural Order being. The spa exposed me to various alternative medical services, such as acupuncture, meditation, massage, and other Eastern medical techniques. I was impressed to find people's health could be improved by some of these alternative therapies, and I felt a positive energy in the ancient Eastern-based philosophies of health that appeared to be lacking in modern Western medicine, the former associating health with the wholeness of the entire human life, including powers that are yet to be well-described by science but seem to exist.

Most appealing of all was the connection between health and the natural world in Eastern traditions. Many in the West believe that a reliance on the natural world for health results from a lack of technological sophistication, but the Eastern philosophy is simply a different way of perceiving these issues, one that has very deep cultural roots. In these ancient traditions, health is regarded as the superior function found within the individual as a whole rather than a defined state of biochemical measurements achieved through technology. In fact, this concept of a whole person, of a person as a complete organic entity within his given circumstances, is greatly lacking in the American healthcare industry. Instead, patients are broken down into smaller and smaller parts as the reductionist whims of science churn out increasingly precise interventions. No doubt there is power to be gained at the microscopic level, but it comes at the expense of the real essence, and whole context, of the person.

I often wonder if it is possible to change a patient's life course and to thereby dispel the patient's disease so that he might regain the biologic function that he was born with (albeit that this innate capacity has been thus far unrealized). This thought has become increasingly poignant, as I am more aware of my health and the relationship it has with my behaviors. I am convinced that there is an unrealized health potential within all of us that will ultimately only be found through consideration of the wholeness of a person. The wholeness we lack stems from the fact that we do not live our lives from the inside (Natural Order) out but rather from the outside (Industrial Order) in. It would also do us well to consider our lives from the context of the bottom (body) to the top (mind). These concepts will be explained fully in subsequent pages.

PHYSICAL MEMORIES

When tending to sick patients, I have noticed that our tissues have amazing recall. Over time our actions change our physical characteristics, and not just a little but a great deal. As I pointed out in "Moments Matter," activities such as eating and exercise induce biochemical reactions that in fact have lasting effects on our bodies and our minds. When I was a child, I heard people say, "A moment on your lips but forever on your hips," which turns out to be true in a very big way. Ultimately, the more permanent effects of biochemical reactions will alter our tissues, our biological substrate that is the context within which we experience our lives.

Thus, humans form many memories, but it is not simply the ones stored inside our brains that impact the quality of our lives. I refer to the memories formed as a result of our nervous systems and primarily stored within the confines of our minds as central memories, and I distinguish these from a different form of memory, one that does not solely affect the nervous system but is stored within our peripheral tissues.

Peripheral memories are formed by organs such as the heart muscle, skeletal muscles, liver, and immune system. It is easy to see that our muscle tissues change with routine physical exercise, both anatomically and physiologically. The physical stresses applied to the muscles cause changes on a cellular level that are not much different from the changes in our brain anatomy and physiology with central nervous system memories. Muscles change in anticipation of future events, similar to the nervous system, and provide an improved performance the next time they are called to duty. Such peripheral memories are strengthened by repetition and intensity, much like memories laid down by our nervous tissue. When we look at body builders, we see the lasting memories of all the bicep curls they have done. The fact that their muscles visibly hypertrophied makes it obvious that memories were indeed formed. As we train in physical fitness and practice our movements, it is the multidimensional effort of the nervous tissue, the muscle tissue, the tendons and the ligaments that provide good recall of the events. We can mold our bodies as we see fit.

Muscles, like neurons in the brain, rely on electrical impulses to do their work. Electrical impulses are generated across cell membranes by ionic gradients and provide the mechanism for the muscle to lengthen or shorten, thereby performing work. Electrical impulses across cell membranes transmit incitement to action from our brain to our muscles and back again. Our muscles are constantly feeding the information of their position back to our brain, which of course puts it all into proper context. All of this communication back and forth occurs in milliseconds.

Peripheral memories occur in virtually all of the other organs in our bodies, and during all activities, but these memories are below the attention threshold for the average person. Ingestion of a donut, croissant, or candy creates a memory within our arteries; and we are in essence training our body to accept these toxic elements as routine encounters in our lives. With increasing frequency, we are seeing liver failure as a result of people eating manufactured foods, high fat meals, and processed sugars. These foods cause fatty deposits in the liver that can lead to cirrhosis of the liver, just like that seen in alcoholics. Our routine meals and exercise activities seem to be innocuous moments in our busy lives, but it is by virtue of these moments that we are creating biological memories in our tissues that have lasting effects.

Vital people have stored abundant memories of vital moments experienced in their lives, peripheral memories that enable our organs and muscles to rally to physical demands. These memories are the essence of physical adaptation to improve ourselves and become stronger. Atrophy of muscles and of the brain is the memorable impression left by disuse. Inadequate exposure to challenging mental and physical circumstances causes pruning of our tissues for the sake of efficiency. That is, our bodies will curtail capacity, tailoring our physical forms to match the demands placed on them. In other words, we practice atrophy just as we practice performance enhancement, and whereas doing more enables us to do more, doing less makes us capable of doing less.

Peripheral memories are also formed in other parts of our bodies, such as the hypothalamus, adrenal glands, and autonomic nervous system. Even liver function improves with each physical challenge. In fact, a more efficient liver results in improved glycogen storage, fatty acid metabolism, hormone production, and kidney communication, so that future stresses and workloads can be responded to. Because the liver can remember previous runs and workouts, tighter control of sugar and fatty acid metabolism is combined with enhanced clearing of the various toxins released by hard-working tissues. Repetitive exercise also improves the coordination of the lungs and heart due to regulatory mechanisms altered by hormones, prostaglandins, and vascular compliance changes. Several glands that secrete hormones change our blood flow and the tone of the arterial walls to regulate regional blood flow, just so we can run around the block. Physical stresses, such as those applied to our bones as we run, walk, and jump, lay the biologic matrix and mineral deposition to prevent bone fractures in the future. Stated simply, vigor is the biological memory of a vigorous life. Indeed, we cannot achieve vigor any other way.

Our immune system also has excellent recall capacity. Interestingly, the immune system needs not only a memory of intruders and threats to which to respond but also is required to remember the self, us, our bodies. Normal tissue can be attacked when the recall mechanism is dysfunctional or when the proper immune response is forgotten and reaches the level of chronic inflammation. Diseases such as lupus and rheumatoid arthritis occur as a result of deficient memory of our own tissue, thereby allowing immunologic attack to destroy healthy tissues. Disordered immune regulation is thought to occur more commonly now than in times past because we as a society are deficient in our exposure to the natural world. Autism, a brain disorder, is now speculated to be a result of the deficiency of the Natural Order in our lives, causing a devastating immunity attack on the child's brain. Activities such as eating and physical exercise govern the basic nature of our immune systems.

A person's ability to accumulate unhealthy memories is very apparent in the hospital. Most patients may believe that their ailments such as a heart attack, kidney failure, or stroke are sudden events, but in reality they are the culmination of a string of unhealthy moments collected over decades that are simply being remembered. But we need not go to the local hospital to see the evidence of stored memories within the entire body. If we can honestly look at ourselves and assess our physical performance, our functional capacity, we can judge whether or not our physical memories are unhealthy.

In biology, static states are illusions. Subtly, we become what we practice over time, becoming more atrophied with less activity and more vital with more activity, for example. However, we modern humans manage to convince ourselves that our biological selves are staying the same when nothing could be further from the truth, which is in fact a very large problem for the individual organism. When I contemplate eating a donut, brownie, fried chicken, or manufactured foods in general, I know what happens internally because I remember what a fatty liver looks like on a CT scan or what an obese person looks like when the surgical knife opens him up. Indeed, if you could see into an obese person cut open, it would be quite obvious to you that our tissues become the processed fat and sugar, the mush, that we put into our mouths. We all choose to make either positive or negative memories (there are no middle-ground memories) throughout our entire bodies and throughout the course of our lives.

Central (stored in the brain) and peripheral (stored in tissues outside the brain) memories interact and depend on one another. In fact, our completeness as an organism depends on a well-functioning nervous system that directs our actions, even those assigned to unseen processes such as immunologic function. Peripheral tissues are recruited and coordinated by the nervous system to engage with life and enhance our performance. For instance, when I run, my brain (central) and leg muscles (peripheral) are coordinating activity based on my desired goals. Since I run often, there is improved communication between the muscles and brain, which together decide on things like muscle recruitment, stride, and tempo. In other words, previous runs induce biochemical memories that are translated into anatomic memories in organs such as the liver, adrenal glands, lungs, and heart. These, in turn, determine my ability on the current run. Today, I remain mindful of the biological context laid down as I perform each run or workout, knowing that later these will allow me to enjoy the next climb up a hill or run in the beach sand. I often state to my patients that, no matter what they have heard about the positive effects of exercise, it is still underrated.

Like exercise, I consider sound sleep to be a necessary component of self-preservation, but it is also important to link exercise to sleep because the one will always enhance the other. People almost universally underestimate the importance of adequate physical exercise to sleep hygiene, instead assuming any sleep issues are the result of stress alone. I often tell my patients that they should not expect to sleep well if they are not routinely exerting themselves physically. The brain functions in such a way that being physically fatigued, not just emotionally tired, at the end of the day is extremely vital to our sleep patterns. Physical fatigue maintains the appropriate physical context for day and night within our bodies. In other words, daily exercise unites our bodies to the diurnal pattern of the planet physically.

I like to think of the day's activities as the wear-and-tear that is necessary to stimulate growth and positive adaptations. The micro-tears in my muscles, the stress placed on my glycogen stores in my liver during runs or the physical demands for better oxygen delivery into my lungs and tissues are recalled in such a way that they become the body's guide to positive changes. Stressful wear and tear by day becomes the architectural matrix for adaptations and improved tissues by night. Organisms with little or no brains still sleep, and therefore sleep is really a product of the body, only made more efficient by a quiet brain, especially a healthy brain. We should consider sleep to be an important time for the body and our health rather than a mental holiday or respite.

Preparation for sleep results from the pineal gland's secretion of melatonin, and while we are asleep, our organs are engaged in "housekeeping" chores that are below the level of consciousness. Growth hormone and sex hormone levels spike at night so that tissue repairs can occur. The guidance for such repairs are our peripheral memories of the day's events that just transpired. This is how we adapt as physical creatures. Good sleep is not just important to our moods and cognitive function but also essential to adaptation and proper function of our entire body. Vital functions such as hunger-satiety hormone balance (described in detail later), growth, reproductive health and our metabolism in general are directed in sleep. Therefore chronic insomnia, night shifts and sleep apnea affect our ability to adapt, heal and be productive people.

Before the dawn, hypothalamic connections within our nervous system (the leader) recruit the adrenal glands and other organ systems, in fact readying them ahead of time to perform optimally at the break of dawn. As we wake in the morning, most of our systems have already prepared for the day and there are increased levels of epinephrine, cortisol and other activating hormones from our adrenal glands. We can stand upright, right out of bed, because our vascular system is already being stimulated by our brain and adrenal glands, the means by which our hearts and arteries bring us to attention with increased blood flow, oxygen and nourishment. This is why most heart attacks occur in the early morning hours, because the brain and body are positioning themselves for the workload of the day even before the workload begins. The body of a person ready to have a heart attack is not up to the challenge of the day. During sleep, we are subconsciously building ourselves up and preparing our bodies for the purpose of the day, and part of that purpose is preparing our bodies for a healthy life.

Sleep is anything but passive. Just as the planet continues its course around the sun without pause, our bodies stay actively engaged in life, whether our cognitive faculties are engaged or not. Sleep is another important aspect of our lives that is often corrupted as we align ourselves with the Industrial Order, again posing an existential threat because poor sleep habits generally reflect a brain that is very concerned about its prospects within the Industrial Order (relationship with technology, career and society) and much less concerned about its prospects within the Natural Order (relationship with body).

Such self-imposed existential threats are common when we consider the mind and body as separate entities, a duality in existence and experience. I would argue that the lack of self-preservation and the roots of self-harm that so dominate our Western lifestyle begin with the idea that the body is not only a separate entity but of secondary importance to the mind. Later I provide ample evidence to dispel any delusion of a separate mind-body existence.

Americans tend to underestimate the importance of the body to the mind just as they underestimate their connection to the Natural Order and a biologically-based existence. People say things like: although I am overweight, I will eat this pizza or this cake because I deserve it, or it does not matter, or we are with company, or I really need the comfort it provides right now. I have said these things myself. Chemical addictions, obesity, and a sedentary life are sustained because the person's desire for comfort matters more than the brain's function (diseases such as Alzheimer's and strokes also occur as a result of such actions, of course) or the health of the body. Why else would people continue to eat breads and sweets while their feet are being amputated as a result of diabetes?

There is indeed much in common between the diabetic who continues to eat industrially produced foods and the alcoholic who continues to imbibe despite the suffering. Although we typically make light of our food choices, the impact this line of thinking has on the suffering that occurs in our country is no laughing matter. Now that two-thirds of Americans are either overweight or obese and a large portion of the other one-third have ailments such as chronic inflammatory conditions, tobacco-related diseases, and cancers that are linked to the ingestion of manufactured foods, the dichotomization of the mind and the body becomes extremely poignant, not only from a monetary but also existential point of view.

Below our level of consciousness is coordination and communication between various organs of the body that result in extremely complex memories, the internal context of our present lives. Past successes are stored with the intent to be recalled for enhanced performance later, and conversely we also know that poor self-esteem, poor health habits, and mental stress wear on our immunity, our performance, and even our entire bodies. I am certain that positive actions utilizing my muscles, liver, lungs, and other organs induce positive memories in my brain at the same time. There is a seamless nature to the effects of our lifestyles, of our choices, on our bodies.

Self-care, such as performing intense exercise, creates positive biological memories inside my cerebral cortex, but a positive disposition is transmitted to the rest of my body as well. The positive feelings cultivated by exercise are greater than the already reinforcing sense of well-being obtained from the associated enkephalin-mediated euphoria. The average American fails to fully appreciate how fitness creates a positive physical context for life that affects all aspects of life. Routine struggles and concerns are put into a different perspective when I feel well, which serves to validate that the most important context for my life is created inside of me.

Watching people at all-you-can-eat buffets and at the concession stands at movie theaters, theme parks, and sports parks is interesting because I wonder if they are aware that these moments will be remembered. They may argue that they are having a grand time, but subconsciously a different memory is being formed, one that threatens their cellular integrity. These people's reality is different from that of someone concerned with optimal physical function and speaks to the power and influence that our Industrial Order culture has over us.

However, rather than dwell on pessimism as a physician who encounters patients with serious infirmities on a daily basis, I choose optimism because I know firsthand the influence our actions have on our lives. I have realized that, if I can use this information to improve my choices, others can do so as well. If they live by a healthy set of ideals that are congruent with the Natural Order, if they engage in vital activities and eat foods that are not industrially produced, people can create a healthy internal context from which to interact with the world around them. When our ideals for a healthy life are aligned to the Natural Order, we can be more certain as we navigate through our surroundings full of enticing artificiality that will corrupt our cells and our minds.

The old adage "we are what we eat" contains more wisdom than most people realize, and I would amend this assertion as follows: we are the memories we make. The biological memories made so whimsically at the dinner table and in the buffet lines, at spectator events and cocktail parties must be recognized for what they are: the moments that govern the essential microscopic flow of forces of our life. In order to sustain an indefinite string of the best biological memories possible, we must respond reflexively to the artifacts posing as food in our surroundings, rejecting them in favor of food from the Natural Order. We must also live like what we are, physical beings that deserve to be active.

Because I have come to recognize that building a better life is born of building better cells, I have aligned my own life with the Natural Order, paying close attention to my biology and the outside influences that might affect it. If the biochemical reactions within my cells are diluted by artificial forces, acting on me internally or externally, I cannot build a better body. I tell my patients that the first act of self-preservation is to recall and preserve the exquisite biology we were born with, for stored within each of us is a physical reality consisting of fantastic natural potential.

THE EXTERNAL CIRCUMSTANCES

As noted previously, modern cultures have evolved through their dependence upon the Industrial Order, and so our reliance upon Industrial Order thinking is perhaps understandable. However, we are first and foremost creatures of the Natural Order, which is not just a substrate from which the Industrial Order arose, but the very context of our evolution as a species and the primary context for our lived existence as individual organisms. We belong to the Natural Order within us more than we belong to our external circumstances, particularly those of the Industrial Order.

We have lost a great deal of reverence for the Natural Order, however, in part because we are insulated from the natural world by our Industrial Order inventions. Thus we are not exposed to and certainly not considered to be part of nature at a conscious level, unless we plan to do so on a vacation to some wild place. Nevertheless, nature lives on within us and we live on within it. Although most of this discussion lends itself to philosophy, the practical nature of this discussion should not be underestimated. I routinely see that the struggles my patients have with regard to their health, which tend to be extremely serious struggles such as chronic disability and even premature death, can almost always be traced to the competition between the Natural Order and the Industrial Order in their lives — and the latter is winning. As modern humans, we become accustomed to eating artificial foods and remaining sedentary throughout our lives, circumstances set forth by the Industrial Order presented to us as unwritten rules as soon as we are born. Being born into modern culture, especially in America, we are expected to participate in this order of operations unless our parents prompt us to do otherwise.

As a physician who specializes in internal medicine, I routinely see the negative consequences, such as diabetes, heart attacks, strokes, and cancers, attributable to lifestyles not aligned with the principles of the Natural Order. Indeed, it could be said that nature is speaking to us, warning us of our gross violations of the Natural Order's rules for life. We are simply not listening, however, as witnessed by the fact that I have a full-time job applying remedies to patients engaged in Industrial Order lifestyles. Their inability to hear what nature is telling them is often fatal. Conversely, I also see the positive consequences provided by nature when we nurture ourselves, eating natural whole foods and engaging in regular strenuous physical activities; but the obverse is also true, that there is much I do not see that serves as proof too. It is rare, for example, that I encounter a person in the hospital who lives by the positive ideals and actions I have put forth in this book, for a leaner, stronger, more efficient body results in, and is a result of, a vital life. Health, happiness, and longevity are not really within the purview of the Industrial Order, obviously, no matter the propaganda to the contrary. Health, like happiness, comes from within us.

Indeed, self-preservation, the protection of our intrinsic value from external threats, is not merely a side effect of, but represents an obligation under, the Natural Order. This notion, like all else that pertains to the Natural Order, is present within us, but it has been corrupted by Industrial Order thinking, like all other fundamental aspects of our lives. That is, self-preservation is often not genuine, as in the preservation of one's self, but transmuted as a form of pharmaceutical and procedural intervention.

The direct source of people's suffering is easily ignored as we fulfill and react to our compulsion to seek a technological, an unnatural, solution to our suffering. Modern humans have taken tool use to such an extreme that we often feel more comfortable with technology supplanting a lifestyle aligned to nature, our very biology, but these are external circumstances created by our culture that adversely affect our health.

Although I am surrounded by the Industrial Order, I need neither be enmeshed within it nor internalize it. I can still manage a resilient stance born of a more conscious relationship to the biological world. This is not easy, however. Once we have determined our external circumstances to be a considerable risk, the only two options available are to flee from Industrial Order society or to build an internal strength with which to withstand the force of the external influences acting on us. We can sustain incredible power when we realize that self-preservation is simply preserving the Natural Order inside of us.

Nature intended health and self-preservation to be automatic, both arising spontaneously from well-functioning tissues. However, in our modern world, it seems that self-preservation has ceased to be so much a drive as a choice, much like our muffins, donuts, chips, cars, houses, or TV shows. Many people seem to say to themselves, "Today, I will disengage from self-preservation because there are better things to do with my time. Maybe after the holidays or my next birthday I will begin to care about myself." Patients often tell me that there is no time or money for engaging in healthy lifestyles or even thinking healthy thoughts (listening to nature, smelling the roses, meditating to secure serenity). They tell me that to live by healthy means is so complicated, that there is too much to do and think about to live a healthy life. My response to these patients is, "If you think health is tedious, try sickness."

Consumerism, including that attributable to the medical industry, can very easily separate us from ourselves, largely because it perpetuates a relationship that has detrimental results for our health. Consumerism of the medical variety also has many abstract qualities that are difficult to grasp, all those numbers that are supposed to add up to health or sickness and all those pills with mysterious effects on our bodies. That is, we measure our health by extrinsic means, such as monitoring various laboratory numbers and ingesting multitudes of pills. In fact, it is important to understand how modern culture in general and modern medicine in particular can corrupt us for our consumerist tendencies represent potent environmental issues that prevent true self-preservation.

I have been practicing medicine long enough to realize that the limited ability of our science to actually deliver health is notable. Very often, Americans may possess a great deal of personal data regarding their physical condition as a result of lab tests, CT scan results, MRI scans, bone-density test results, and the official blessings from the medical industry they receive in doctor visits; and yet the data does not often translate into real health. In fact, despite incredible gains in our knowledge about the human condition, health remains rather elusive in America. Perhaps our accumulated knowledge and Industrial Order focus on our "vitals" have not delivered as promised because of the concomitant failure to foster self-care and reverence for the Natural Order.

Analysis of the numbers and intellectual musings regarding biological processes do not foster a natural appreciation of our physical condition. As we acquire interest in the manmade things surrounding us, we tend to lose interest and respect for our bodily processes that give us life. Our bodies do not matter as much because functional loss is quickly accommodated for and the loss seemingly controlled through the use of innovations, or so modern humans think. We seem to have the capacity to remain comfortable in a very abstract world and believe it is a healthy place. This appears to be a weakness that allows us to intellectualize our way into disease. Indeed, Americans are sold the idea that health is an intellectual endeavor, a function of problem solving only, and perhaps this is part of our failure. Self-preservation is in fact a "mindless" activity, as evidenced by its presence in the most primitive life forms.

The application of medical knowledge may improve our chances with regard to sickness, wherein there is actually a problem to solve, but there is no problem to be solved when it comes to the notion of health. Solving problems and placing Band Aids is different from being a champion of one's fate, a healthy state wherein issues are resolved by the intrinsic strength of the person, and there is more spontaneous well-being all around. Therefore an approach to health that focuses on problems rather than the beautiful spontaneity of a natural life not only lacks credibility, it is certain to make people disillusioned. The real problem with most people's health is that they are not working with the Natural Order to secure the health and fitness that is intended for their lives, the spontaneous health that arises so easily for the vast majority of the planet's animals.

If we pause to consider our external world and the recommendations our society throws at us for our so-called health, we may realize much of it lacks common sense. For instance, it is difficult to fathom that aggressively sedating and then inserting a 120-centimeter fiberoptic camera (colonoscope) through the rectums of the masses of people who are experiencing no symptoms of disease would actually improve the health of a population of 300 million people. Similarly, the chances of health seem unlikely if society's plan entails extracting blood from tens of millions of fingers several times per day in an effort to warn people of blood sugar anomalies. Nor would we likely improve the health of a large population of people by systematically concealing muzzles inside their abdomen.

Arguments for most of the medical interventions on offer may seem concrete (evidence-based, they say), but are actually quite abstract when we consider that they are machinations, usually based on evidence that cannot be guaranteed even with repeat testing. Making medical science and the so-called health it delivers even more abstract is that other people have been studied, not the person signing up for these tests or treatments. We are biological organisms and very real, yet we place faith in abstract concepts as if they were real, as in real enough to secure our health, as in superior to real things like good old-fashioned diet and exercise.

Medical systems tend to busy people with all sorts of minutiae, technological gadgets as well as facts, much of it not at all related to their health. My patients tend to fixate on medical minutiae, even when I try to persuade them to consider their lives as whole entities that need not be relegated to the sheets of paper, the MRI and CT scan results. I tend to encourage the view that their lives have physical qualities that they can test on their own, through experiments in diet and exercise, but this often evokes silence, perhaps a blank stare, and then followed with more references to the medical data sheets. It is not healthy to allow the abstract world to override the real world.

As if it were by grand design, the medical community's abstract intimacy with people's organs almost exactly mirrors the lack of physical intimacy patients tend to have with their organs. It is indeed curious that our most potent methods for assuring health, those with time-tested survival advantages (diet and exercise), will often be promptly abandoned in lieu of something new that is generally a meme, a modicum of information, or an abstract thought, such as a cholesterol molecule that can rule the fate of one's life. Knowledge and the micromanagement of small aspects of one's life lack real power and sincerity because knowledge tends to be applied in isolation and outside the actual context of one's life.

Modern medical techniques have themselves become a lifestyle, a lifestyle advocated as the consumer's best choice for health. The fast-paced evolution of this lifestyle has indeed resulted in phenomenal complexity, culminating in the "patient-centered medical home," one of the ultimate acquisitions of an industrial identity, wherein the distinction between man and machine, the natural versus the artificial life, become blurry. Within the patient-centered medical home, the person is likely to forget that self-preservation is about them, and not the machines sent by industry. It is easy to see a lost distinction between a person's life and the life of a machine in such a vision of how we should live, our cells, and therefore our lives, increasingly aligned to the intentions of a very unnatural order.

Such increasing alignment, due in part to the quickening evolution of medical technology and in part to the profit motive of the medical-industrial complex, means these are trying times indeed. To make matters worse, increasingly Americans are becoming sick, and at the very same time they are being pummeled with potent pieces of information. It seems that our society's lack of health and physical realism creates not only physical sickness but also an insatiable desire to consume more information. In short, Americans are out of control. Their bodies, and their notions of their bodies, are becoming distorted, and they are becoming people they do not wish to be; and so they attempt to regain control of their lives through the acquisition of information, which, sadly, increases their mental corruption as regards health.

The information age allows for the fabrication of new medical information with ever increasing speed, and then allows it to be disseminated — after first being enhanced with sensationalism — with viral fervor. In short, we are increasingly at risk for being duped by those purveying unwise, often misleading information that is frequently based on well-contrived scientific evidence. As a result of the sheer quantity of this information, and the fact that it can't all possibly be true at once since much of it is contradictory, I suspect that we will be increasingly dependent upon our personal instincts in the future and perhaps upon a renewed sense of physical realism, knowing health when we see it and feel it. Medical experts and medical panels are being hatched with incredible speed, as the Industrial Order demands, and eventually we will achieve a homeostatic state, an equilibrium of sorts, within the matrix of medical information. There will be multiple expert panel positions for every possible medical consideration, and thus, given the obvious absence of consensus, hopefully individuals will take responsibility for their own health in a more realistic way. Instead of assuming everything they read is true, they will realize the built-in contradictions and will have no choice but to believe what they see in the mirror and how they feel.

This dilemma of contradictory expert advice is apparent right now. For example, some experts say that we should prescribe no more cholesterol drugs (recent data suggests they harm women disproportionately) and other experts say that we should give them to children. We have experts who say there are too many mammograms performed while other experts say too few are being performed. We can find experts who want to run PSA tests and operate on prostates while at the very same moment other experts say that the federal "war on cancer" is too inclusive. Indeed, despite the on-going war on cancer, it is increasingly more common to hear medical experts recommending that we can live with some cancers, even some forms of breast cancer, because a worse fate than the disease is in store for those pursuing treatment.

With regard to people's weight, there have been some recently disturbing and contradictory news as well. The January 2, 2013, issue of *JAMA* recently published data from researchers K. M. Flegal, PhD, and B. K. Kit, MD, that showed being overweight is associated with longevity, and more incredibly that sustaining Grade I obesity (women with approximately 45% body fat content and men approximately 33% body fat content based on a BMI of 35) did not shorten one's life. These researchers imply that being overweight is healthier and being obese is okay. Should we try those numbers on for size if we want to live longer? With time, if it is not obvious already, we will see that medical science is more concerned with its interests than our interests. The reporting of this study is likely to hurt more people than it will help. Furthermore, what about credibility in general? Why should we place our faith in a system that claims to be "evidence-based" when its evidence is so convoluted and changing faster than weather patterns?

There is of course speculation as to why these so-called "excellent" studies discussed above could draw a contradictory conclusion to the prevailing wisdom, saying now that being fat is better. Most of this speculation centers around the idea that overweight and obese people tend to receive more medical care. But this brings us to another implication: that medical care creates longevity, and therefore we are better off being medically dependent, a fact that is not proven. Using reductionist methods once again, this *JAMA* study is unlikely to hold truth for our personal lives and we invite great risk applying their conclusions to our lives. Very important but absent in these medical studies is the concept of an authentic, non-medically-dependent life, one that will only be known to the individual and never revealed to the medical researcher. For my reader, the most important thing to remember is the notion of sincerity and authenticity for life. We risk a great deal of our health and quality of life when we place our faith in the hands of people and systems who do not really know us.

I am not always reassured by the abundance of medical data and the peer-reviewed scientific evidence gathered through rigorously controlled medical investigative trials for an obvious reason: these studies are often more apt to promote corporate health than human health. An abundance of information may make us feel intellectually reassured, our human appetite for information satiated, but then we are also at risk of being overwhelmed and paralyzed by indecision. I often sense that a person's mental processing centers are working hard to understand the myriad medical concepts and minute details of his condition, meanwhile nothing is being done, only considered and calculated by the individual. Furthermore, an unwavering plenitude of information automatically lacks sufficient context in the life of the individual considering it. Many of the newest truths being delivered to us are also likely to invoke unnecessary fear so that our attention is effectively diverted from older truths, such as health is a product of eating right and being active, not drugs and artificial interventions. Fear can be a useful ploy when there are contending truths, a ploy to get one to err on the side of caution and purchase a given intervention, but this strategy obviously places one's health at risk because health decisions based on fear are rarely wise decisions. Lastly, having too much information thrust upon us, much of it contradictory and playing upon our fears, tends to create disillusionment, which injures our spirit for life. Ultimately, mankind cannot live by health information alone. We must cultivate our physical reality within our biological moments and our biological memories.

THE BUSINESS OF SYMPTOMS AND FEAR

Americans complain of many symptoms which can be a sign of dysfunction and potential disease, but they are more likely to be just feelings or mild alerts. A great deal of medical resources are used, not on treatments for the sick, but on evaluating complaints such as palpitations, joint aches, chest pains, heartburn, fatigue, and insomnia. The resources are spent expeditiously because simple symptoms translate very efficiently into tests and procedures, even when there is no disease. Thus, symptoms can themselves be harmful, not because we have medical conditions associated with them but because they often result in unnecessary radiation exposure, invasive procedures and medications that can cause complications. Accurate symptom identification is necessary for self-preservation in general, but in this day and age perhaps even more so.

Symptoms will come and go in our lives, and the initial evaluation will be made by the person having the symptom, hence the biological memories and physical context we have built for ourselves is extremely valuable. The better our context for evaluating a symptom, the better off we will be inside and outside the medical arena. Although chest pain can be a sign of a heart attack or a blood clot to the lung (pulmonary embolus), most chest pain is not lethal and can be caused by a number of things such as anxiety, muscle strains, arthritis, heartburn (GERD), viruses like the common cold, and pleurisy. There are two obvious reasons why we are prone to think the worst for ourselves. One, most Americans are engaged in self-harming lifestyle behaviors, therefore lacking a positive physical context from which to evaluate their symptoms. Two, we are culturally conditioned to investigate our symptoms and fear the worst. I routinely encounter people receiving expensive and potentially harmful evaluations for these two reasons, and the sad part is that the reassurance provided during the extensive medical evaluation tends to have an extremely short half-life.

If a person does not spend sufficient time and energy getting to know herself as a biological entity, sufficient time being mindful of her physical performance and engaged in self-care, sufficient time simply paying attention to her health, then she will be far more likely to process any symptoms of pain or fatigue or other symptoms through a context of fear. In fact, I sense that many of my patients believe that it is only a matter of time before their anticipated cancer, diabetes, high blood pressure, or heart condition will present itself. It is difficult to convince them that their chest pain or abdominal pain is benign because they almost unwaveringly expect that a serious medical infirmity is associated with their symptoms. Deep down they know they mistreat themselves, so the worst outcome should be expected — at least until they present for medical care when they always expect good outcomes will be guaranteed. There is an undercurrent of fear in America and the medical industry both cultivates it and preys on it.

Lipitor ads say, "It was a horrible feeling. I couldn't believe I was having a heart attack." A Bayer commercial portrays a patient saying after his heart attack, "I don't ever want to feel that powerless again. . . I will trust Bayer." When the advertisements promoting pharmaceutical products emphasize "doing enough," they mean ingesting enough of their pills, of course, and they are preying upon our fears to maintain a large consumer base. Many aspects of modern medical care are engaged in predatory behavior, especially with respect to the cultivation of fear.

We see advertisements on billboards that portray colonoscopies as life-saving tests, but gastroenterologists fail to mention that a low-fat, high fiber diet and regular exercise save more lives from colon cancer than do colonoscopies, let alone that such personal interventions also reduce the burden of all of the other chronic diseases. Undiagnosed hypertension is broadcast as a silent killer, and there has been an explosion of hypertension experts and new advertisements for that. However, these experts don't tell us that most people do not need to get hypertension in the first place, and, if they do have it, many people can eradicate it themselves.

When we are afraid, the threshold for the application of technology and spending money on medical interventions is lowered. Fear prompts us to turn over our faith and resources to technologies that promise to deliver us from the disease we dread, but the presence of fear means we lose control of our ability to make thoughtful, informed decisions. The medical industry is in fact quite adept at making us "worried sick," as University of North Carolina professor Nortin Hadler, MD, so eloquently titled his book.

Fear itself causes disease too. I previously noted the role of excess screening tests in furthering disease, such as mammograms inaccurately diagnosing cancer, but the tests, particularly false positives (and up to 25% of mammograms are re-tests based on faulty original mammograms), also create stress that negatively impacts the immune system, making infections and cancers more likely and causing heart attacks and strokes. In our medicalized society, we are consistently reminded of our malignant tendencies that induce a special mindfulness or "cancer awareness," but these reminders are devised to instill just enough fear to undergo screening tests rather than change our lifestyle enough to improve our odds of not developing cancer. Generalized fear or disease awareness does not promote wellness nor is it necessary for self-preservation.

As we understand less about ourselves in a physically functional way, we will reflexively fear more (fear of the unknown). Therefore any symptom we have must be placed in the context of a biological organism, the context of how we care for ourselves and what we know about ourselves. For example, a fifty-, sixty-, or even seventy-year-old person who is lean, does not smoke, and vigorously exercises almost daily is able to evaluate his chest pain in a more positive light than a much younger, obese, sedentary person who smokes. As I was writing this chapter, an advertisement for the local Chest Pain Unit crossed the television. It said, "If you have chest pain, call 911 now!" I was actually experiencing pretty severe chest pain at the time because I had worked hard on my chest muscles in the days prior to this. I am over forty-five years of age, and therefore, the logic goes, this could be heart related. Or it could be that I strained my muscles digging holes in my yard or that this pain is secondary to a recent viral infection. I am not likely to assign my pain to heart disease, and not because I am a doctor (on the contrary, doctors "know" anything can happen), but because I know I am a healthy person and confident about my health. Moreover, because I am in tune with my health, an engaged participant in my self-care, I know that it is highly unlikely that I have a heart issue, and I tend to have enough visceral awareness to know when I am in trouble. I have built a positive context into my life, one born from the Natural Order and intrinsic to all of us. We can all sense our health status in meaningful ways on a subconscious level (this mechanism will be explained more fully in the next section).

Symptom evaluation, such as chest pain, is big business for medical centers because fear is a potent instigator for a person to seek investigation, especially those people who know very little about their own potential for disease versus health. Across the US, there is a new proliferation of accredited Chest Pain Centers, specialized units within hospitals dedicated for evaluation of the one symptom: chest pain. These specialized units sound encouraging and garner good media attention and other notoriety by being certified by prestigious accrediting agencies such as the association of Chest Pain Centers. The name of the centers would lead one to believe that lives will be saved, but in actuality, the same evaluation and management of chest pain has been done and is done by less designated centers, such as one's local hospital devoid of a Chest Pain Unit. Indeed, the principle behind the name appears to be a "selling" point rather than a therapeutic point for patients. Oddly, only people who are at low-risk for having heart disease are admitted to the unit. The screening procedure for entry into the specialized Chest Pain Unit begins in the ER where patients have been determined to have a low risk for a cardiac condition. In other words, a patient must be deemed extremely unlikely to have heart disease to the extent that it will immediately take their life before they can be admitted to the specialized "symptom" unit. High risk patients experiencing chest pain are not accepted in these specialized Chest Pain Units, they are believed to have heart disease that threatens their immediate life.

Why the need to evaluate patients that have exceedingly low risk of heart disease? The centers provide the hospital with a new source of revenue, plain and simple. Hospitals have merely found a way to apply the same extensive cardiac work-up to more patients, albeit younger patients and healthier patients who have exceedingly low risk of heart disease, thereby expanding the market for their services. These Chest Pain Units are quite expensive to operate, and so they must churn through hundreds of healthy patients with only symptoms and no heart disease in order to justify their creation. Can we justify this if there is already an ER, a specialized cardiac unit, and even a cardiac wing of the hospital equipped to evaluate chest pains that might or might not be an indication of a need for urgent care? Indeed, the evaluation of every symptom of chest pain not only costs time and resources but also exposes the patient to harms and potential harms that will be delineated below. Like all aspects of medicine, all tests, all procedures, and all medications, we should always ask ourselves: are we helping more people than we are hurting? I wonder because a Chest Pain Unit really amounts to a better way to attract low-risk patients for evaluation at a particular center (designated by an accrediting agency), as well as provide an easier pathway for people unlikely to have immediately threatening heart disease into the hospital to spend a day or night. Many people who enter a Chest Pain Unit will leave with the same chest pain that led to entry and not know what caused it. The less we can sense about our health in a natural way, the more harm can potentially be done to us by entering the medical system. I encourage all people to know enough about their health in a visceral sensory way (through the practice of being healthy) so that they can process their symptoms meaningfully, thereby minimize the need for entry into the medical-industrial complex just for symptom evaluation.

One day, I approached a thirty-eight year old patient who had been admitted to the hospital for chest pain. The patient was a bright, young lady with a happy disposition and appeared healthy by the average American standards. In fact, she did not appear to be unhealthy in any way. She, like my other patients of the day in the unit, had experienced an ill-defined chest discomfort and, lacking sufficient internal knowledge of her health status, became concerned enough to proceed to the emergency room. When the cause of her chest pain was not determined during the initial evaluation by an ER physician and other medical staff, she was admitted to the Chest Pain Unit (a specialized hospital area for the evaluation of one particular symptom) overnight and underwent the usual battery of tests, those being two more physician evaluations, serial blood enzyme analyses, serial EKG's, a chest x-ray, a CT scan with a heavy intravenous dye load, and a stress test employing a physician assistant and an echocardiogram machine (ultrasound test). This ordeal which was initiated by a simple symptom resulted in medication exposures and radiation to her body (CXR and CT scan) and generated thousands of dollars of billable medical care. For the hospital, the value of this encounter is rather concrete and quantifiable in dollars and cents, but the value to her and society is more abstract.

I arrived at her bedside with good news: all of the tests were normal. As I pulled back the curtains of the divided room, she was eating a large donut with icing and colorful sprinkles. I told her that this did not seem to be appropriate behavior for someone admitted to the hospital with concerns about her heart. She laughed and promised she would change but proceeded to eat the donut in front of me as I explained the tests to her. Her biological moments did not seem to matter to her. This young lady obviously did not believe that her moments spent in the Chest Pain Unit were important aspects of her life or that her donut was a serious contribution to future potential illness in her life. This apparently healthy lady, who had a husband and kids at home who depended on her, did not appear to understand her ability to direct her own health fate. Health for her was an abstract concept as it is with most Americans; the Chest Pain Unit simply represented an intellectual comfort zone (is it really comforting?), where we can be cradled by industry for a day or so and our fears addressed. Perhaps we can all benefit from more sincerity with regard to our physical realities rather than more medical care.

BUILDING PHYSICAL REALISM

Virtually any American need not have the health he or she has now. We build our internal physical realities one biological moment, one biological memory at a time; and a better life can be realized when our biological moments matter and the biological memories matter. Simple, fundamental activities that will improve our health require virtually no thought, no intellectual justification or explanation. Instead of seizing upon these simple activities, however, because of our rigid industrial identities, we are convinced that health can be achieved through the industrial prowess of the medical industry. The patient in the above vignette who did not appreciate her physical reality, her intrinsic biological value as a human, despite spending almost twenty-four hours in the hospital with medical attention and the intellectual deployment of sophisticated technology, is in fact the Industrial Order norm. Americans generally enter the medical machines with little sincerity or realism for health and exit these machines with the same amount or less. We can be extensively processed by modern medical techniques and through various standards of care with little health ever being changed for the better; preservation of the Industrial Order occurs while self-preservation is never considered. And this is unfortunate because, as we spend so much of our resources on maintenance of the medical machinery, we have fewer resources for the truly good life associated with the preservation of our natural selves.

In times past, a human's authentic physical nature was the norm and a healthy body was essential to a good life. Although our ancestors were more inclined to consider their vital possessions their livers, hearts, and lungs, we consider manmade artifacts as our vital possessions and the measurements of our success. However, we remain biological beings, and through our connection to physical nature we can remain the experts on our own health. It merely takes re-realizing that it is the manner in which we live our lives that preserves the value of our lives.

Health is indeed valuable and we should be inspired to have it. But in American society, the value of our health is mostly considered as it relates to the medical costs accrued by illness, and, as for inspiration, we tend to find that in pictures of healthy people on a cereal box picture or a magazine cover. Under these circumstances, we have ignored the true value and inspiration inherent to a healthy life, and this imposes a more negative context to the average American life.

Excellent biological memories, those creating a solid foundation for inspiration and a positive context for life, build us up. There is a great deal of value and inspiration to look outside at nature while considering the possibilities for my physical life today. My body is ready for just about anything; my only limits are generally imposed by the Industrial Order with regard to time. The concept of time is important to health and therefore we will not avoid it in our consideration of physical realism.

The average American has a very distorted appreciation for time. For instance, the words "I would take care of myself better if I had time" are spoken often. As I hear them spoken in the exam rooms, I stop my patients to inform them that all they are guaranteed in life is time. No one is guaranteed a house, job, money, a functioning government, security or anything else — but they are guaranteed time during their lives, all the way to the time they die. Our lives simply unfold physically as we fulfill the obligation of time. Lack of a physical reality, such as the reality of time, has huge implications for our lives, especially for our health.

To be engaged in our health is to remain connected to the sum of our biological existence — our moments and memories that provide us our physical reality, including but not limited to those created within our liver, fat tissue, muscles, nervous system, blood vessels and immunological system. These tissues create our presence in and allow us to participate effectively in this world. The brain interprets the data arriving from the body's senses and organizes the responses. A better functioning body not only creates a better presence (certainly healthier), but also transmits better data to the brain. Simply put, a better functioning body conveys more accurate information and therefore gives rise to a clearer reality.

Oxygen obtained through the lungs is pumped by the heart and delivered to the brain by the intricate pathways of the arterial system. Sugar that fuels brain function is supplied from the gut digestion of the foods eaten, its regulation and storage in the liver, and its entry into the neurons by the insulin produced by the pancreas. These routine activities we take for granted are sensed by the brain, being information that represents our status with regard to life. The simple act of carrying too much fat impedes oxygen (sleep apnea) and sugar utilization (diabetes), meaning that our cognitive realities as created by the brain are altered — but not for the better, as obesity and diabetes create impaired realities that become hardened with time (later called dementias). It is the physical memories of our present circumstances that provide the foundation for our sense of reality during our entire lives — our physical presence now having great impact on our future sense of reality. This fact should inspire us to be healthier right now.

Before we became so numb to our own physical existence, health was a genuine *feeling* of wellness, one that could even provide us inspiration, in that we were required to be well to be a success. We can all be inspired today by the feeling of wellness itself, unless we remain numb to our own performance. Furthermore, before we were cognitively hijacked by modern medicine, health was not an abstract concept, in that it was not defined as the meticulous management of chronic disease, the definition many of us subscribe to now. There was only one method to health, and it was both physical and natural. Today we have that choice, but we also have the choice to believe in a more abstract form of health, such as health delivered through the medical industrial complex, a health that is represented by symbols and numbers that are surrogate makers of normalcy. The abstractness of the medical-industrial definition of health should not be underestimated because the surrogate markers we tend to monitor are actually most useful to signal disease, and we should not necessarily feel or believe we are healthy simply because our surrogate markers are in a normal range ("the clean bill of health"). Believing such a prospect represents a breakdown of reality.

Mindful of these circumstances, it is practical, if not imperative, to focus on improving our lives through physical means and considering the inspirational value of real health through the Natural Order. It is once again notable that only humans, who alone have the power of intellection and command of the abstract world, have an intrinsic problem with health. This is important to recognize if we desire real health. We need not intellectualize about self-preservation because it is a given; therefore if we lack it, we simply lack physical realism, perhaps insanely trapped in an abstract world. Health is indeed very physical and found freely in the less complex animals, therefore it is unlikely to require our intellectual capacity to secure health or acknowledge our health. In fact, we are more likely to think our way out of health (proven through the Industrial Order) rather than think our way into health.

People who are physically fit know that they are fit because they experience fitness throughout their bodies, not in an intellectual manner. Physical fitness provides satisfaction, and this can be transmitted to extrinsic aspects of our lives, such as better work, better play, and improved relationships with other people. In effect, we can experience a better life through a better body, the sensory prism and connection of our minds to the world. The information that makes physically-fit people feel well is personally-derived, intrinsic knowledge rather than extrinsic knowledge or data garnered on a laboratory print-out or from a doctor visit. A simple diet, resembling that of ancient man, and the sweat of our brows can bring us the most sincere form of a life, can create intrinsic value in our lives. Implicit to the notion of self-preservation is that we are familiar with our true self, the self we intend to preserve. Will we preserve the self as created by the Natural Order, or will it be the self as constructed by the Industrial Order? Only through physical realism will we know the difference.

Another way of saying this is that the manner in which we live our lives is what provides the physical, intellectual, and emotional context to our lives. We need not be preoccupied with medical minutiae, stuff that appears more whimsical than solid evidence these days, to provide the context for our health. We know when we are taking care of ourselves because we feel the effects. We also know when we do not take care of ourselves, and the negative thought content that results from this is likely to alter our perceptions of ourselves and of the entire world. This is important since most people take better care of the things they value, therefore it is important to have a body that *feels* valuable. A body aligned to the Natural Order is more valuable to one's mind (literally) and this creates a positive reality all around. Interestingly, each of us has the ability to enhance his physical value and his physical awareness.

Interoception (achieved via the interoceptive pathways, distinct anatomical structures in our brains) is the awareness of our entire body, the awareness of our vital systems and the activities of life that make us vital. This most valuable information generally lies below our consciousness unless we are adept at utilizing it, as trained athletes manage to do routinely. Survival skills are enhanced through interoception because we grow more acutely aware of our most vital systems and how we relate to our physical structures, our organs.

Indeed, interoceptive wisdom can only come from physical training because it is our body's relationship with our physical activities and the choices we make that change our physical nature. We tend to become wiser about ourselves as we become more flexible and stronger. We can fine tune our internal sensory networks and increase the capacity to understand ourselves, but, perhaps more important, our minds can be taught respect for the body when it is more valuable to the mind. In a few words, interoceptive wisdom is derived from the experience of being healthy.

Many people fail to understand that they build their own negative physical realities as much as they fail at the positive ones. This makes incredible sense in the scheme of a biological life because, as it turns out, there is no static state: we are either building our tissues or we are allowing them to degrade. Likewise, we are either building a stronger, more vital physical context for our lives or we are building a negative one (either by negative actions or by inaction, which yields atrophy). In other words, we are either training to be champions of our health (living a Natural Order existence) or we are training to be patients (living an Industrial Order existence).

This training results in the context from within which we interpret symptoms, as mentioned previously. A person who has had a heart attack will interpret a pain in his chest differently from a person who has never had a heart attack, but the concept of biological memories forming the context of our lives is yet more poignant in this example. After a heart attack, a person is more likely to assume that any sudden chest pain is another heart event, even if it would have likely been dismissed as non-cardiac pain prior to the heart attack. The context of the same chest pain arising in the same person has been forever changed because his biological memories have changed. His interceptive wisdom has changed and now fear is also at play. Visceral memories run deeper than cognitive memories.

This perceptual change occurs not just in the brain but also in the person's endocrine system, peripheral nervous system, and vascular system. Stress hormones are typically released from the adrenal glands prior to someone even beginning to be concerned about a heart attack. The body feels the concern first and then alerts the brain, but the context has already been largely determined and therefore the actions associated with this symptom are already in progress. In fact, heart patients, especially those with failing hearts, will always have excessively activated stress hormones because this system is forever on guard after a heart attack, awaiting the next sudden threat to life. What we eat and what we do with regard to physical performance is how we alter this reality. Modern medicine tells us we must call 911 immediately, which will draw us into the labyrinth of the medical industry, unhealthy or not, if we heed that advice. The only way to effectively counteract these potent external forces is to have an excellent internal physical reality wherein a chest pain is not likely the sign of a physiological cataclysm.

This is not to suggest that symptoms, indeed the vitals that the medical industry is so fond of tracking these days, should not be heeded. I am merely suggesting that heeding them need not amount to industrial medical interventions. Indeed, my vital signs can be very useful, serving as guides toward positive habits or at least as prompts to eradicate negative habits. If my weight goes up, I need to change my routines, and this is what self-preservation is about: preserving the real self, the one before the weight gain and chronic disease. Like my performance in exercise, my blood pressure and cholesterol are also part of my attributes, indicators of how I care for myself. Within the auspices of modern medicine, we are inclined to interpret our vital signs as problems but do not connect them to the activities that create the vital signs.

Medical intervention tends to amount to suppressing those vital signs, which then cease to be barometers of our lives and a measure of the efficacy of our lifestyles.

Suppression of our vital signs without acknowledging the role of our behavior in those vitals is a clear indication that our culture mandates a disconnect from actual biological reality in favor of an Industrial Order version. Our physical reality is replaced by a treatment plan. Treatments not only have their limitations, some of which will be elucidated in later sections, but treatments also change the context of our lives, one not aligned to the fantastic optimism found in the Natural Order version.

Physicians are mandated, literally, to place every heart attack patient on beta-blocker drugs because the goal is to subdue the constant hyperactivity of the endocrinologically mediated stress system while blunting the patient's heart rate and blood pressure. In effect, physicians thereby change the internal milieu of the patient, the internal biological context from within which this patient lives the rest of his or her days. Beta-blockers typically cause adverse side effects, such as apathy, erectile dysfunction, loss of libido, diabetes, decreased exercise tolerance, and even depression. Beta-blockers are not the only method of disease maintenance therapy after a heart attack, however. Cholesterol drugs, aspirin, and other blood pressure medications, called angiotensin-converting enzyme inhibitors, will provide additional control over the patient's internal milieu. People undergoing such treatments remain blunted throughout their bodies in order to suppress their vital signs, but their vital signs are not the problem.

There are eminently practical reasons for living aligned to the Natural Order in that remaining outside the medical industry can improve our survival rate. The negative consequences from one's lack of self-care are now compounded by the potential for the negative effects of being a patient in the medical system. A recent article in the American Medical Association's newsmagazine, *American Medical News* (April 25, 2011), stated, "One-third of hospital patients experience adverse events and about 7% are harmed permanently or die" after admission. Indeed, every medical test and procedure has error rates and complications, no matter how prestigious the medical center and physician. In 2009, 2.1 million hospital-acquired infections resulted in

77,000 deaths. These risks are made exponentially worse inasmuch as patients often leave the hospital with a strong potential for bad outcomes and the potential need for more dependence on medicine. The only way for a patient to prevent this is to use the hospital only if absolutely necessary. Instead, I see people use the medical system as if it were the safest place on Earth, oblivious to the threatening realities and side effects of simply entering the medical arena. Perhaps we should consider the risk-to-benefit ratio as we enter.

One would think that, with a second chance at living a healthier life, people would vigorously activate their drive for self-preservation after having suffered a heart attack, stroke, or newly diagnosed chronic medical condition. Unfortunately, in my experience this is typically not the case. In fact, almost none of my patients appear to activate their drive for self-preservation when faced with illness, and I find this particularly disturbing. The patients' lifestyle behaviors are almost never addressed meaningfully, despite the preponderance of evidence stating that these behaviors are causing their health problems and in spite of their own negative experiences within the industrial healthcare system. The upshot appears to be that a healthy context must be built from within long before the person reaches the prestigious medical system, for, once embarked upon the treatment path, people seem all the more unlikely to change their lives to avoid that path.

Furthermore, when people aligned to the Natural Order become patients, their ability to confidently assess their symptoms from within the physical context they have achieved is all the more valuable. For exceptional healing and health to take place, the context through which a person lives his life is an essential element, the proverbial glue, that unites the abstract medical markers of health to the natural being, the patient. This contextual glue consists of the patient's insights, perceptions, desires, expectations, beliefs, feelings, and, finally, his physical work at being well. These crucial elements of a human life must be well established and, I would argue, made physically real in the form of how the person lived prior to entry into the medical arena. Otherwise, the myriad of Industrial Order artificialities become all the more urgent and important and consume the consumer, as it were. The sharper the focus on the health within the patient, the more likely real health can be achieved under all circumstances.

If I am having difficulty piecing together a problem, in order to attempt to put the physical issue into the context of the patient's life, I will often ask the patient, "What do you think is wrong with you?" I have received very interesting feedback from that question. Some patients will thank me for asking. Others appear offended by the question and respond with something like, "You are the doctor, and so you should know." The first response indicates an appreciation for the human-to-human contact needed before I can make sense of the available medical information. The latter response generally tells me the person has unrealistic expectations of the medical system, viewing it as their health savior via technological intervention. Absent in the person providing the second response is the consideration that my tests actually rely on his insight and ability to relate his symptoms to me in a focused, meaningful manner. If he cannot provide me reliable information about himself, it is difficult for me to understand anything about him, including his health. With abundant, seemingly unlimited resources at our disposal in our medical suites, physicians are compelled to order more tests and procedures in the hope that the lack of biological/life context found in the patient will ultimately be compensated for by a series of tests and consequent medical treatments.

It is unfortunate when this situation happens, because the more tests I run and the more specialists I consult regarding the patient's condition, the more information that needs to be placed into patient context. More information is not necessarily better information, and in fact it often causes a complex patient presentation to be more confusing. We end up trudging down the path of chasing test and scan results with more tests and scan results. Each result carries additional false positive and false negative rates, which means we can quickly confuse ourselves and the patient — but the patient is stuck with the side effects (and cost to society). Patients are best served by physicians when they present to the medical conglomerate with a sound internal awareness of their health status, either through intuition or knowledge of their functional capacity.

A recent study from the July 2010 *Annals of Internal Medicine* showed that the risk to patients without their own context of health is great indeed. This study by Weiner, Schwartz and others stated: "We found high error rates among physicians who were confronted with clinical situations that required attentiveness to a patient's context." The context for medical care is not exactly the same as our context for health, but it is clear that, without a firm identity of health, a patient's individuality will surely be lost when he enters the medical arena and that loss is dangerous.

Navigating through the increasingly complex maze of potential tests and therapies will surely be more difficult in the future because there appears to be little chance for a reduction of the systematic approach to healthcare, wherein we are treated relative to the scheme of a standardized system. Worse, inside sterile medical centers are several third-party entities with a vested interest in the person's sickness, making the patient's situation relative to treatment all the more confusing, without the patient ever knowing what the multiple sources of that confusion are. When a patient arrives at the medical center or the doctor's office, there are standards of care issues mandated from the government and medical societies, the physician's personal preferences and business interests, pharmaceutical marketing interests, device manufacturer interests, infrastructure interests (clinics and hospitals are concerned with staffing pressures and system-directed patient flow, now referred to as throughput), software interests, revenue interests, lawyer interests, insurance compliance interests, and government regulatory and financial interests (very deep bureaucracies) that are all brought to bear with each and every patient encounter, no matter how small the physical complaint that initiated the visit. When this many interests from so many third-party agents are involved from the moment of patient contact, the perspective of any encounter is greatly distorted, and, as always, the patient is the one paying the price, both literally and with his health.

Therefore, each of us should be determined to preserve our true self and avoid the intrusion of third-party interests, the abstractness of modern medicine and the technologies that usually further separate us from ourselves. We deserve the best context for our lives, that being the most personal context that we build through physical realism.

SELF-DETERMINATION

I repeat: the true health of a person is not reflected in tests or in a medical diagnosis made by an expert but is instead reflected in that person's performance. A large study performed through Stanford University in the 1990s evaluated whether or not long-distance running led to more incidence of arthritis in runners compared to the general population. The researchers periodically took x-rays of people who did not run and people who did run, and they found that both groups had the same incidence and degree of arthritic changes in knee x-rays over time. However, the runners differed from the non-runners in that they rated their arthritis as less noticeable and much less likely to limit their activities. Even people who ran long distances into seventy years of age had fewer complaints of arthritis, despite x-rays confirming its presence. So, what use is the diagnosis or x-ray finding of arthritis if a person's performance provides a better assessment of his joint health than an x-ray test?

Another example: back pain is a common problem in America that consumes tremendous resources in medical costs and lost wages. Studies have been performed to analyze the incidence of arthritic or diseased backs in the general population using sensitive MRI scans. Comparisons were made between people who complained of back pain and those who did not. The frequency of diseased backs, as diagnosed by MRI, turned out to be the same for both groups. In other words, the appearance on the MRI scan had nothing to do with the symptoms or even the disease. Our backs are obviously what we make of them, not what the radiographs and scans make of them.

Likewise, recent studies with regard to Medicare-funded treatments of lower-back pain provided similar insights. In this study, an examination of records revealed that complaints of back pain are evaluated by MRI in certain regions of the country more than in others. In those regions where more MRIs were performed less physical therapy was prescribed before surgery. The regions that performed more MRIs performed more surgeries, but the pain and disability outcomes were worse. So, the conclusion is obvious: patients need to know, for self-preservation reasons, that to assure a better outcome, less pain, and disability in the future, they should avoid MRIs and surgery unless absolutely necessary (which should be the patient's determination).

My father, who is eighty years old, had a recent physician encounter wherein his perspective of his own health context became very important. He visited his orthopedic surgeon to complain that his foot with a fallen arch was giving him pain. My dad's foot is in fact grossly deformed, but he accommodates his malady enough with orthotics to mow and trim and plant large trees in his immaculate yard, the maintenance of which is his purpose in retirement. The doctor ordered x-rays of the foot as well as his spine, which the doctor told him revealed that his spine was a "mess" and that he was at risk of having paralysis at any moment. He referred my father to an orthopedist who performs only back surgeries.

My father explained to the original surgeon that he has no problems with his back and never has back pain. In fact, besides working in his yard chopping trees, planting trees, moving shrubs, and digging large holes with shovels and pick-axes daily, my father still does about 250 sit-ups at the gym on an incline bench almost every day of the week, just as he has done since my childhood. I can testify that my father has excellent core musculature and no disability from his back, but the surgeon said he would not operate on his foot unless his back was fixed because, according to the x-ray, it was that bad.

After his orthopedic office visit, my dad went to the gym. While at the gym he could not help thinking about the purported frailty of his lower back as he was doing his sit-ups. Was he going to be paralyzed by doing these exercises? Maybe he should limit them or give them up all together as the expert had led him to believe he might become paralyzed. The expert physician scared him, and so he later called me with these questions.

This is a perfect example of how tests and physicians provide no concept of the real health of the person. Only my father knows what his back health is. More important in this anecdote is the fact that the surgeon has caused potential harm in my father's health because he led him to believe that his vigorous lifestyle might be putting him at risk for paralysis. Many people would find it difficult to block out entirely what the specialist, the bone expert, has said about the bones that encase their spinal cord.

To put my father's issues in perspective, which is the notion of self-preservation in action: he has done everything in life to care for his back, and that is what is most important to his back. His core musculature is strong and that will determine his back's fate more than what an x-ray or MRI scan reveals. He has prepared himself to have a good fate with his spine despite the physician's assertion and x-ray results to the contrary.

In fact, even random health crises will have a better outcome if we are prepared to have a good outcome before the crisis. For instance, if two people are subjected to the same exact car wreck in an experiment, the health outcomes can be very different, even at the same medical center and with the same team of personnel attending to them. The person who is in excellent physical shape will be more likely to survive and will have less pain and disability compared to an unhealthy person. After the hospital rescue, the person's potential for excellent rehabilitation is far less dependent on the rehabilitation facility but instead on the patient's level of health before the injury and the patient's desire for rehabilitation.

The equation seems a no-brainer: greater health before the auto accident equals greater health after it. Additionally, however, the well-rested person (not suffering from obesity-related sleep apnea), the physically toned person, and the person with an uncorrupted nervous system (from the effects of sugars and diabetes) will have better reflexes and could potentially avoid the accident in the first place. Trauma units are grand inventions, but producing safe and physically fit drivers is a sounder approach to trauma care. Moreover, I would argue that fewer car accidents occur when people respect their own lives sufficiently not to be distracted by text messages or intoxicating drugs.

In the scheme of our biological lives, the power needed to recover from all adverse events begins long before the adverse event occurs. The power a person has with regard to his own prevention of disease and trauma should not be underestimated. We may not be able to stay safe throughout our lives, but we can certainly suffer less if we have physical and mental resiliency earned through self-improvement. Self-care strengthens our minds and bodies and thereby prevents fatigue, depression, cancer, heart disease, diabetes, strokes, and virtually all other chronic diseases. Self-preservation is really self-respect.

AN ETHIC OF SELF-PRESERVATION

Self-preservation as defined in these pages is the act of preserving the life we have been given, but the life we have been given can take many physical and mental forms, as determined mostly by our choices. Therefore our physical form and our physical function are the result of how we live. I am fortunate to have discovered a meaningful lifestyle that has changed me in body, mind, and spirit.

No matter how much scientists wish to promote the notion that physicians wield incredible power over the lives of their patients, I still marvel at the healing that is unexplained. My favorite moments as a physician are those in which I can educate, motivate, and empower my patients to take care of themselves. There is no explanation or science for that particular method of healing, and yet it is the best method as regards outcomes. There is a force within each of us that can be tapped into when we are not so corrupted by the Industrial Order, the same force that takes care of the rest of life on Earth, the Natural Order. This force of life provides us the universal means for self-preservation: self-assembly, self-awareness and self-determination. Preservation of the self is more aligned to a physical reality than to an intellectual method of life.

There is a great deal of discussion regarding the conservation of our planetary resources, but the same people don't seem to care about self-preservation. Intellectually, the conservation of our bodies can at least be compared to the conservation of the planet, but in fact the preservation of the natural resources innate to our bodies is as important as any other planetary resource. Indeed, given the Industrial Order connection to the devastation of both, saving the planet and saving our own health are very much akin.

It is not unusual to witness people discussing the need to convert to energy-efficient cars or light bulbs as they dine on Doritos, Pepsi, donuts and Whopper sandwiches, all of which are associated with very large factory complexes. While conversing about sophisticated environmental concerns, perhaps reviewing the benefits of eating organic locally grown foods, many of these same folks are awaiting the precise hour to take their cholesterol medication, blood pressure pills, diabetes pills, and sleeping medicine — all of these also made in large factories. Does it really matter whether our medications are made in foreign countries or locally or that the medical complexes have become so numerous that they are deemed "local" in spite of being part of some huge chain of such complexes? Shall we convert these medical complexes, which consume a great amount of resources and run 24 hours a day, to nuclear power or wind generated energy? Should the paramedics' vehicles and helicopters run on natural gas or biofuels? How we treat ourselves not only speaks volumes about how we treat the natural world, the two are intimately connected. Aldo Leopold once said (as quoted by Edward O. Wilson) that the first rule for successful tinkering with ecological and biological systems is to keep all of the natural parts as they are. There is much less to remedy in the world if we stop its destruction, and that includes the destruction of ourselves.

At my favorite Mexican restaurant the other night, I asked for no tortilla chips with my order, but the young lady went ahead and gave me the usual serving that was called for with each meal. After we finished eating, I felt excellent, but I noticed the enormous pile of chips (several hundred calories worth, and the worst kind). I sat there with my wife and thought about the tortilla chips that I did not eat. Of course, I chose not to eat them because I do not like the way they affect my body, but why did I not just eat them the way I used to, automatically, as it were, because they came with the meal?

Although I know how bad they are for health reasons, these chips do not repulse me as a putrid piece of meat would. They seem at the least innocuous, and the salt was visible and I like salt. This is what is different about our modern circumstances. Abundant foods that can make us quite ill do not repel us like rotting meat, soured milk, and other toxins. Therein lies the element of choice that is our health situation in the modern world, healthful living as choice rather than a drive. There are many reasons to eat the tortilla chips, such as I already paid for them, there are billions of hungry people in the world, the people around me may notice and think me wasteful, they used to taste good and probably would again if I routinely ate them — the list could go on. But there is only one reason I should not eat the tortilla chips: my life and the preservation of an excellent biological moment, one without the industrial-made internal chaos that results from consuming such things.

The drive for self-preservation is still strong in us when we are faced with threats that elicit mortal fear or particular scents that repulse us on an instinctual level, like the smell of rotting tissues. But when threats are made occult (even presumed beneficial to us), disguised as something they are not (like food), there is no deep-seated drive for self-preservation elicited, and indeed the lack thereof is made worse by cultural acceptance of these products. We are vulnerable because our biology has not evolved to handle these well-disguised threats, some that even hijack our intentions for life.

In times past, physical insults to our bodies that were detrimental to our overall health were delivered by physical blows, and the notion of self-preservation was more concrete. Today, we are routinely assaulted by seemingly friendly forces. Food appears friendly enough, for example, but our foods today can kill us despite the words of encouragement like Nature Made, organic, Breakfast of Champions, or a silly cartoon tiger saying they are great. It simply takes time to do the damage, but therein lies the seduction that leads to the altered perception of the members of Industrial Order society. We are also coerced into engaging in invasive medical procedures and into consuming chronic medications by people who "care" for us. Sickness is a physical insult, especially chronic sickness, but we are cajoled into adopting the chronic disease lifestyle through words such as wellness and health. In fact, quite often the best options for one's health may not be discussed by those entrusted with our health.

Physical harm need not even be derived directly from an extrinsic threat. Many of my patient encounters confirm the perception that society has undergone an ominous transformation in its thinking. I must climb an enormous mountain of trust to recommend that a patient not engage in further testing or discontinue medications or avoid an invasive procedure that may require general anesthesia. The patient often interprets these suggestions as a lack of concern or lack of "care" on my part, whereas too many tests and too many procedures are almost never considered a lack of care or concern. Obviously, self-preservation is an art in our age, one that requires cultivation of our critical thinking skills initially and of a thick skin in the face of skepticism. The ambiguities within our society are best countered with a strong dose of physical reality, the reality we create within us.

There is no ambiguity as regards a healthy life for any organism except humans. Curiously, even unicellular organisms exercise self-preservation, and yet they cannot be ascribed an intention to such behavior because they have no nervous system. All of a unicellular organism's life activities are geared to preserving itself. Thus, it is safe to assume that the mind of man is the main obstacle to a healthy lifestyle for humans, the mind that is first led astray into believing that the food industry actually produces food and that the medical-industrial complex dispenses health, the mind that becomes sufficiently corrupted to stand by as the body, the vehicle for the life the mind serves, is also corrupted, unto death. We can all do better than that. By being aligned to the Natural Order, we improve our internal and external circumstances, creating a position of strength as we relate to the Industrial Order. To that end, I have undergone an authentic biological adaptation through and through to assure the best life possible, one that will be described in great detail. This adaptation is very real, not at all abstract, and can be seen and felt in my new identity: I am a Naturvore.

THE NATURVORE

IDENTITY

Naturvore: 1) an organism that does not consume manufactured foods, instead relying solely on natural, whole food for nourishment; 2) a person with a strong ethic of self-preservation who internalizes the principles of the Natural Order; 3) a person who interacts with nature to improve his physical condition.

Like most Americans, I became overweight in middle age, but my story is a bit different in that I acquired poor health despite completing years of training in health care, becoming board certified in internal medicine, and after practicing modern medicine for more than a decade. It is most curious that I was considered knowledgeable and competent to treat my patients with regard to their health, even as I lost my own health. Eventually, I became uncomfortable in my own skin and finally realized that I was deficient, not in intellect, but in a particular form of wisdom: nature's wisdom.

Because the methods to change me could not be found within the healthcare industry, I was forced to look outside the traditional medical establishment. Out of desperation, I simply began to act like a healthier person, and as I achieved a healthier state, I increasingly understood the wisdom of the Natural Order. In short, my very essence has changed over the last few years.

In 2005 and early 2006, I performed substantial research into the plausibility of food being our source of optimal health and potentially even our medicine (Hippocrates is quoted as saying this). I researched various super-fruits as promoted through multi-level marketing schemes but found more promises than substance. I also investigated the therapeutic potential of homeopathy (credible to many Europeans), received body massages, and attempted meditation as a practice. My single most important act was replacing manmade foods with naturally created whole foods; I did not do this all at once, but nevertheless the results were extremely positive within a short period of time, and this was most enlightening. As I consistently sought the nutrition from nature, I had coincidently lessened my exposure to manufactured foods, and, in hindsight, this was more valuable. My body quickly began to change for the better, and by the summer of 2006, at forty-two years of age, I concluded that manufactured foods were indeed a threat to my health, and therefore I stripped them from my diet completely. By virtue of rather simple actions, I began to see incredible results, which coalesced into a belief system and finally an identity. I became a Naturvore, although I had yet to process this new self-awareness.

Since then, my nutritional requirements have been met by whole natural foods alone (or, as my brother-in-law says, food from the ground). With improved physical function due to a new method of living, I acquired a newfound respect for the Natural Order. I did not grow up this way, nor was I formally educated to think in this manner, but the positive results I have achieved were, and remain, indisputable. Shortly after converting to a Naturvore, persistent aches I was having in my neck, legs, and feet disappeared, a result of the fact that I am prone to much less inflammation on this new diet.

Physiologically, my aging has essentially reversed. My weight dropped precipitously, and I am consistently below 10% body fat. My pulse and blood pressure reside around 60 and 110/60 respectively, while my body temperature is also down to below 97 degrees much of the time. As "age-related limitations" vanished, I became more confident in my own skin, so to speak. In fact, I find myself performing physical feats more comfortably and competently than when

I was twenty years old. For example, after turning forty-two, I ran a six-minute mile and completed a 13.1-mile run in one hour and forty-five minutes. These physical performance improvements allowed me to engage in more fun physical adventures, and I began to run distances for enjoyment, while reveling in my connection to nature more and more.

With a diet devoid of manufactured food and with regular exercise habits, my physical attributes underwent significant change. In effect my physical identity was changing for the better as it was more physically defined by nature. Of course, I felt more alive, with improved physical prowess. A new pattern of living was taking shape in my life as my mind and body achieved congruence. Therefore I became confident that I would engage in this way of life for a long time, and I felt compelled to give my new identity a name. So the "Naturvore" identity was coined on a bright, warm summer day while driving somewhere between Fruit Heights and West Point, Utah, with my family.

We were on our way to celebrate a friend's birthday, I felt really good about the exercise I had done that day and was quite pleased with my new eating plan. As I drove, I realized that I did not wish to negatively alter the metabolic advantage I had achieved through exercise and proper diet just to engage in a social function, a backyard birthday party. Even though my family was bringing salad to the party, I knew there would be the potential for an awkward social moment when someone offered me cake or chips. My intentions for the afternoon were clear in my mind: I did not wish to corrupt the excellent biologic moments I had established within my cells. I was still recovering from an intensive workout earlier in the day, and so the last thing I wanted to do was to eat non-nutritional foods and disrupt appropriate healing.

I wondered, during awkward social moments such as these, what reason I could give for refusing to indulge in the traditional birthday celebration foods. The answer: a catchy descriptive name could impart the seriousness of adhering to a belief system. A similar response is provided by vegetarians, who, when offered meat, say, "No, thank you. I am a vegetarian." It is a philosophically defensible position. I decided that I would be referred to as a Naturvore. This seemed ideal because the word itself specifies the particular foods I will eat and, conversely, the foods I will avoid, and therefore there would be little ambiguity for those I told of my new identity. More importantly, the Naturvore identity solidified my essence.

While living in Utah, I often ran through farmlands, and I have given much thought to the animals locked in pens on those farms. They too are trapped in the Industrial Order. They too are not free to seek life as nature intended. Instead, they are dependent on man's inventions and delivery of produced goods to sustain their lives. These horses and cows share a particular kinship with me in this regard: intrusion of the Industrial Order into our lives. Fortunately, I am able to escape and run as a man in the wild because I have that choice. I feel most complete as an organism when I am engaged in long runs. Around mile six or seven there is an overwhelming sensation of optimism for the future and a connection to something much larger than me. I am generally more satisfied on my runs than at any endeavor in the civilized world.

Experiences such as these gave me a sense of a new communion with the natural world, and I began to think that just maybe a natural life is what nature intends for all of us. Some of my feelings can be attributed to the endorphin release in my brain during the run, of course, but those too are part of the Natural Order. It is speculated that nature provides such good feelings as a means of positive reinforcement for this most beneficial behavior. But there is more to it than that. In those moments I am one with my biology and fully dependent upon it to get me home. This is a much larger appreciation than that derived from reading and studying about the biochemical pathways, as I have done through the years of my career as a physician. I can sense that my lungs are taking in the necessary amount of oxygen to fuel my metabolic machinery while I adjust the depth and rhythm of my breaths. Indeed, during a run I connect to my organs in the most meaningful way and become present for the moment. I leave the civilized world in those moments and reside only in the natural world that gives rise to my physiology.

Civilized concerns tended to evaporate as I behaved as an organism that belongs to nature. Even objects that were a source of great pride, such as my large brick-and-stone executive home sitting upon the Wasatch Mountains, became less important to my essence on these excursions into the farmlands. During these runs, my success in life was purely aligned to my performance on the run and not aligned to the house or the civilized lifestyle associated with it. I have found a deep-seated feeling of certainty for my future from these runs, certainty that has opened my mind to the prospect that one's future can be best determined by one's physical fitness and need not be dependent upon society's products and schemes.

It is satisfying to know that, if I maintain my health and physical performance, I am much more likely to have a future that is exciting and personally gratifying in a physical sense. I feel great optimism in believing that I will be productive and capable of energetic work and play far into the future. Why shouldn't I, if I continue staying active and youthful? Our nation is in fact facing a future financial situation likely to be less stable that it is now, and therefore our only real security may be in maintaining our health. Indeed, my health may be the most important determinant of my future financial security — and a safer bet than trusting financial planners and the government to provide for my future life. I remain comforted by the biological moments I am experiencing now, those being the best guarantee of my future security and health, thoughts that make this middle-aged man very happy.

After establishing a certain degree of improved health, I was able to isolate the origin of my obese state and the myriad aches and pains I was suffering. The origin of such misery was my acceptance of rampant cellular corruption in my life, based on my acceptance of a physical identity aligned to the Industrial Order. Later, after becoming a Naturvore, I began to consider the usefulness of identities in general. Wild animals never lose their biological identity or their connection with the natural world as humans do. A biological identity is rather simple for animals, not layered with the artificial complexities of a civilized world, of course, but domesticated animals lose their biological identity as well. The simple deduction is, therefore, that the context, our Industrial Order world, to which both humans and domestic animals are subject, is to blame for that loss of a biological identity.

Humans are born with a biological identity, that of *Homo sapiens*, created from our DNA sequences. *Homo* refers to our genus, relating us to the other upright-walking primates known as hominids. The word *sapiens* indicates a particular species of hominid; the Latin translates as "the wise one." That is, we are all two-legged upright-walking animals with relatively large, high-energy consuming brains.

Animals in general can be classified as herbivores, carnivores, or omnivores. Interestingly, humans are not just omnivores in the traditional sense of the word (our diets consisting of both meat and plants). Humans can also be considered omnivores in that our species consumes both artificial (that which is manufactured and does not occur naturally) and natural food products to meet their nutritional requirements.

Within the context of my Naturvore identity, my reasons for exercise changed. I was not exercising to work off a bad meal but instead exercising to live a fuller life, the life I was expected to have as a *Homo sapiens*, a biped with a brain. I was not yo-yoing any longer as if a healthy life was a constant battle, but instead found a steadier course of health, complete with a newfound serenity. I suspect non-human organisms experience an extreme serenity in that they never question their identity or allegiance to such things as the Natural Order. For human beings, awareness of ourselves in relation to the other organisms on the planet is the mark of a conscious state, and we cannot be self-actualized if we remain oblivious to the Natural Order, the ultimate force behind our lives.

When my beliefs and actions remained in congruence with one another, I gained self-respect, and this in turn was positive re-enforcement to act accordingly in the future. It became obvious to me that this identity not only provided me the strength to overcome social pressures and temptations, but also, because an identity guides us much like an internal compass, I was propelled toward desired goals on an unconscious level.

Identity, as defined by Webster's *New World College Dictionary*, is the aggregate of the characteristics and qualities of a person, and regarded as essential to that person's self-awareness. All cultures and most religions foster the formation of an identity that influences thoughts, guides behavior, and provides perspective to a person's life. The essence of a person should be embodied by their identity, and yet humans seem to forget this most important element of their identity, a biological essence. Indeed, aside from physical attributes associated with gender and ethnicity, our identities are rarely thought of in biological terms.

For some cultures, health states and longevity are part of their national or cultural identities. North Karelia, Finland experienced very high mortality rates from cardiovascular diseases during the 1970s, and a physician named Pekka Puska took it upon himself as a local physician and public health servant to improve the health of the members of his community. He espoused a healthier lifestyle and engaged his fellow citizens by challenging and prodding them to act healthier. The people of North Karelia were persuaded to get more exercise, despite cold conditions most of the year, and they followed through. He encouraged people to eat leaner meat, less butter and more vegetables, and to quit smoking. After a few decades, the North Karelians experienced a 73% reduction in cardiovascular disease and a substantial reduction in the number of deaths from cancer.

It is noteworthy in this day of advertised modern technological miracles and fiscal constraints that all of these improved outcomes in population health for the North Karelians were accomplished without any high-tech medical devices or advancements to medical science. People simply lived better, then felt better and so were more productive, and the result was that the positive effects became contagious and enhanced wellness was incorporated into the North Karelian culture.

In 2010, I attended an event graciously put on by a local hospital system, Park Ridge Health, during which Mr. Dan Buetner, a National Geographic contributor, explained that his research team studied four areas of our globe notable for having a higher percentage of centenarians than does the rest of the world. These "Blue Zones" are located in Sardinia, Italy; Okinawa, Japan; Loma Linda, California; and Nicoya, Costa Rica. None of these cultures consumes substantial quantities of prepackaged foods or heavily processed foods, and many inhabitants maintain outdoor gardens and walk a great deal as a necessity. These cultures have diets consisting primarily of natural whole foods and their lifestyles are physically challenging. Overall, in other words, the people's lives are active, natural, and purposeful — in many ways representative of a successful natural existence. Some of the interesting differences among these Blue Zones are that the Okinawans are known to pause before meals and say, "Hara hachi bu," a spiritual request to only eat until their stomachs are 80% full. The Sardinians of Italy perform a significant amount of daily walking and keep close family ties, activities thought to enhance their wellness, and the people of Nicoya, Costa Rica, have a simplified rural lifestyle and are prone to eating smaller meals and performing a significant amount of manual labor to sustain their lives.

The residents of Loma Linda, CA, Buetner studied were Seventh Day Adventists, and they do not engage in the typical American lifestyle. Because of their religious beliefs, they don't smoke or consume alcohol, they eat a vegetarian diet, and they routinely engage in more exercise than the average American. The Adventists are proof that better health outcomes can be achieved by establishing a healthy identity even while living within areas where a low-quality diet of manufactured foods and a sedentary existence are the norm. In fact, the inhabitants of all of these Blue Zones are proof that people, regardless of their wealth or lack thereof, can improve their health prospects if they have a personal identity aligned to healthy habits. Moreover, entire cultures can be transformed to states of health, but the changes must come from within the individual (his beliefs and values), in essence altering his identity sufficiently from the Industrial Order norm to achieve a Natural Order identity. Unfortunately however, the effects of the spread of the Industrial Order can be seen in these Blue Zones, for they are slowly losing their longevity as new generations within their cultures increasingly become influenced by Industrial Order products, including manufactured food.

The biological identity that we are born with, and that is still present within us, is masked by Industrial Order civilization, our biological identities even altered by our civilized identities. For instance, I am an American, a father, a husband, a doctor, and various other titles. These roles can often take precedence over our biological identities, even at the risk of our physical health. In the modern age, we develop behavioral and cognitive attributes, defined and provided by society, that further define an individual's identity. The most intricate and influential human identities tend to be aligned with the roles we undertake as we interact with society, but we still have a choice. As humans, creatures with free will, we can choose to awaken our biological sense of self to incorporate into these other roles, indeed to guide us as we undertake them.

Our goals in health as biological beings could be simply stated. There are three pillars of vitality utilized by all animals in the biosphere to achieve an optimal life and health: 1) remain lean, 2) remain active, and 3) remain connected to nature. These attributes of successful animals in the wild should serve humans well too.

FOODS OF NATURE

Natural whole foods are constructed from the organic world by living organisms; they are most capable of nourishing our cells to achieve the health nature intends us to have. Natural foods are simply that: foods that are eaten in their natural state, such as vegetables, fruits, berries, nuts, seeds, legumes, beans, lean meats, fish, seafood, eggs and dairy. These foods improve our bodies through the delivery of highly bio-available forms of macronutrients (carbohydrates, proteins and fats) as well as micronutrients (vitamins, phytonutrients and antioxidants) that help organize and optimize our bodily functions, therefore making us more resilient to disease. Recall that our bodies are the soil of our lives and therefore made better with natural constituents. A varied natural food diet provides the optimal quantities and combinations of vitamins, minerals, anti-oxidants, and essential fatty acids to sustain us. This is the same mechanism of acquiring nutrients that made us successful biological beings throughout the vast majority of our history.

From a mechanistic point of view, vitamins are meant to be ingested as mixtures in whole foods, not supplements made in factories, because they are more completely absorbed and utilized in the presence of other vitamins, minerals (selenium, magnesium, and phosphate), and even fatty acids. Many of the vitamins we eat are best delivered to the intestinal wall in lipid micelles (fat globules) that provide for their transport across the wall, therefore salads should be eaten with healthy fats. For instance, unsaturated fatty acids from olives, avocado, and nuts promote more complete absorption of the beta-carotene found in sweet potatoes, squash, broccoli, carrots, apricots, and peppers. Certain vitamins are not very active biologically without the presence of others, such as vitamin C assists vitamin E, as well as iron metabolism and so on. Vitamins generally require other nutrients to work, so that taking them in supplements, in isolated forms, tends to diminish their usefulness.

The nutritive contents of fresh fruits and vegetables tend to be housed in specialized biological devices that guarantee the best absorption and assimilation of minerals into our bodies. The specialized delivery devices encasing the genetic material of plants appear to be designed to contribute to the success of the animals eating them in order to ensure excellent results as regards the spreading of the plants' seeds. When ripening, the fruit is more metabolically active, and therefore more alive, than when it is growing prior to ripening. Therefore apples, grapefruits, grapes, berries and others are still alive as we are eating them. Their life energy is transferred to our life, their biological constituents serving to improve ours. Things that have evolved together tend to develop a codependency and relationship that is usually under-appreciated. Simply stated, life serves to propagate other life.

Whole foods such as fruits, vegetables, and nuts can be heated, crushed into mixtures, spreads, compounds, and concentrated forms while not altering the basic substrate provided by nature. These are indeed acceptable sources of nutrition as long as the final product does not contain artificial ingredients and is not drastically different from its original form. Milk, which occurs naturally, is something of an exception in that it should be pasteurized for mass consumption. However, pasteurization does not create a product that is substantially different from the milk when it came out of the cows and goats.

Thus our local supermarkets, farmers markets, and roadside stands are filled with vitamins in their most usable forms, and virtually all corners of modern America have a consistent supply of whole foods throughout the year. Super fruits, whole foods with extraordinary nutrition, don't need to be exotic. For instance, blueberries, cranberries, and pomegranates can be found in most grocery stores. My super-fruit choices are black grapes, grapefruit, apples, mangoes, kiwi, and pears; and I eat just about the whole list every day. The super-vegetables are broccoli, tomatoes (actually a fruit), garlic, onion, beets, spinach, chard, and mushrooms. I suspect people could take a daily dose of fruits and vegetables as easily as their medications.

I did not design this diet, of course, and in fact it was in place long before our Neolithic times, when we moved to grain-based societies, and began to habitually design and manufacture foods. This was the diet of the hunter-gatherers, the human diet that gave rise to the Neolithic man and ushered in the most distinctive modern human attributes. As Harvard anthropologist Dr. Richard Wragham has suggested, the brain volumes of our Paleolithic ancestors grew from 900cc to 1300cc, where it is now, coinciding with (or due to) the hunter-gatherer diet consisting of animal meats, fish, nuts, legumes, and whole fruits and vegetables.

Our current state as a species in the modern world appears less than optimal, and if the species continues on the same path, future generations will be further stripped of vibrancy. The dietary changes associated with our most modern times appear to be the cause of our poor health prognosis. On a more optimistic note however, we can take our cues from our ancestors and make food choices that are likely to improve our health. The historical success of our species is aligned to the nutritional habits of our ancestors, and our tissues still share those strong developmental links. Our evolution took place within and as part of the biomass of earth, a living matrix that created and sustained us, and so it makes sense to replenish our fabric with the fabric of nature. Therefore, we should align ourselves to the nutrition of the past, not the present or the future. I choose natural, whole foods in place of manufactured foods.

MANUFACTURED FOODS

As I relate in my story, we invariably begin our transformation to a more industrial identity as soon as we are weaned from our mother's nipple. Within our first year of life, manufactured foods are generally introduced into us and we begin to change cellularly and morphologically. There is no age at which we are less vulnerable to the detrimental effects, obvious in many of today's children, appearing less healthy, less optimal, and also subjected to more medical care and medications in the prime of their youth.

We accustom ourselves to manufactured foods, and then the effects on the brain set in, and we continue to acquire more of an industrial identity. What I found after decades of cellular corruption was that the impact of our industrial identity corrupts us on a very deep level because it changes our course in life. I suffered a detrimental diversion from the health (and brain activity) I was intended to have. The only way to know our healthiest body weight and our most authentic human form is to remove manufactured foods from our diet and see the results. What I have found in my experience is that the body people have while eating manufactured foods will not be supported by natural foods —

they become much healthier eating natural foods. The percent body fat or body-mass-index (BMI) sustained while eating natural, whole foods is likely to be our healthiest level possible while our nutrient stores are likely to be the most optimal.

My posit is largely clear and need not be certified by a medical institute: the entire epidemic of obesity in America can only be sustained on diets consisting of manufactured foods. There is virtually no way that natural whole foods alone would sustain the obesity epidemic and chronic disease burden in America. Each one of us can try by our own experiments, and I suspect that maintaining excess weight (especially weights greater than 300 pounds) by eating natural, whole foods is virtually impossible. The barriers to becoming obese, and certainly super-obese, are innate to the foods themselves, such as absence of the "sugar high" associated with manufactured foods.

In the United States, a significant portion of each person's daily calories comes from manufactured foods, including soft drinks and juices made with high fructose corn syrup (HFCS), processed/assembled meats (often containing HFCS), candies (HFCS) and other prepackaged grains. These Neolithic innovations were supposed to improve our lives. Bread-like products (such as crackers, cupcakes, tortillas, pastas, chips and other salty or sweet grain products) and corn-based products (including many candies) have created a culture of habitual grazing. Habitual unhealthy eating is mostly associated with products infused with or made from grains. Emotional eating that adjusts moods has obviously become a grave health threat for the majority of Americans, but any correction to this is likely to be met with resistance because there is money to be made the way things are. The bottom line we need to remember for the sake of our health is that food from nature propagates life in many forms, while food from factories propagates industries.

As I removed manufactured foods from my diet, I was forced to more fully understand the nature of our food, manufactured versus natural, and, upon further refinement of my distinctions, I realized that bread is an artificial food. Bread has an aura of being natural but is in fact a manmade food item very removed from its original source. Frequently, we are led to believe something very different as we are exposed to commercials that show a healthy-appearing person standing in natural wheat fields, while the product they are selling, an "all natural" bread or organic snack bar, is actually created to human specifications and assembled by machines in factories. The product consumed is in reality far removed from that picture, not at all resembling the amber fields of grains in which the person is standing. A great deal of industrial innovation and manipulation must be undertaken to fashion grains into a packaged product. Other products that should be considered manufactured foods are muffins, bagels, rolls, and donuts.

The denouncement of bread as a natural food is likely to be met with resistance because of its almost universal use as a staple of the human diet for the last 4,000 years. Bread is also mentioned in many spiritual texts and therefore holds cultural importance through much of human history. Scientifically speaking, bread heralds man's ability to cultivate, store, and manipulate a natural substance into a desired food product. From an anthropological point of view, bread marks man's ability to make his food consistently, separating us from the other animals. Therefore, it can be said that the advent of modern food production in general can be traced back to our success with bread production. Subsequent sophistication in bread manufacturing techniques allowed the fabrication of breakfast cereals, pastries, cakes, granola bars, bagels . . . and the list goes on.

In light of our cultural history linked to bread, I can see why bread is considered a natural food by so many people, but in reality bread is made in factories and wheat itself is not the most ecologically sound crop. Only by virtue of American industrial might are grains (wheat, oats, corn, etc.) and soybeans — crops that would otherwise be extremely scarce (they do not reproduce and grow well without the aid of humans) — able to occupy approximately 200 million acres of the American heartland. For instance, it takes 1,000 tons of water to grow one ton of wheat, which leads to enormous water losses in erosion because grains do not have a deeply penetrating root system. Grains typically deplete the topsoil of nutrients, and therefore huge quantities of fertilizer made from fossil fuels are applied to these fields to sustain production.

I consider bread and bread products as being life-less because the vital life force that is in the wheat used to make bread is stripped from it long before these manufactured products reach the grocery store shelves. Wheat must be harvested and degraded sufficiently before it can serve as the initial manufacturing input for bread production. Refined wheat can also be an input for other factory-made products such as noodles, breads, cakes, and pastries. But before wheat can be made edible or fashioned into these other innovations of foods products, it undergoes intense processing. First, wheat is harvested, then threshed and milled, all by machines such as large combines. These refinement activities also consume substantial amounts of water and fossil fuels. Afterwards, when the wheat is in a manufacturing-grade form, it is milled and mixed with more water and other ingredients (oils and salt) to make dough. This dough is then "proven" by controlling the ambient temperature and humidity to allow yeast to convert carbon into CO_2, thereby creating gas bubbles. Gluten is the glue that allows the bubbles to form and gives bread and pastries that airy quality. The greater the temperature at mixing, the greater the number of bubbles. The dough rises as these gas bubbles are formed. It is estimated that 3% of flour is turned into CO_2; these non–nutritive carbon atoms are released into the air (*Cereals and Cereal Products: Chemistry and Technology*, D.A.V. Dendy and Brogden J. Dobraszczyk).

Further production and manufacturing occurs as the dough is placed into ovens to be baked. Strict hygiene rules are followed, as the warm, moist dough is ripe for bacterial and fungus growth. Fungicides and bactericidal products are important at this stage of production since the equipment must remain sterile. Dough that does not meet taste, texture, and color standards will be promptly discarded. American consumers have specified standards for their processed grain products, so even the color of the product is extremely important. Baking bread, of course, requires further expenditure of fossil fuels to maintain temperatures between 200-475 degrees Fahrenheit. The bread loaves are then sliced by machines, but the shape is specified, and that means a certain percentage is surely discarded due to improper slicing and an unapproved shape. The packaging is typically plastic, derived from fossil fuels, to provide a seal for moisture and maintain a physical barrier. Packaging is performed by machines, and again a substantial use of electrical power is required. The finished loaves are inspected for quality, and some of the product is again discarded because the entire package may not satisfy final inspection before departure from the factory.

When we walk into any supermarket, we are greeted by entire walls of bread products. Entire mini-shops inside the larger stores provide us with substantial grain-based human artifacts, each one a direct descendent of mankind's initial creation that marks our transition to a civilized life. With the exception of the produce section and the butcher shop, although many people consider these industrialized as well, the rest of the store contains thousands of products that have been created as a result of man's success with manufacturing bread.

Large bread displays are important because the consumer believes there will be a fresher, more acceptable loaf somewhere on the wall. Consumers want their bread to have moisture condensed on the packaging and a warm feel when they pick the loaf from the shelf. Baked fresh hourly or daily means just that, and the loaves not sold in that time frame are discarded to bring in the new loaves. American consumers will not tolerate old bread or squished loaves, and so much of the bread produced is never sold. Preservatives are applied to most bread, but that introduces potentially harmful additives to the consumer's diet. If no preservatives are used, then bread either goes to waste more quickly or requires freezing, thus using additional oil-based energy inputs.

Bread does not last long after you bring it home either, and it is estimated that 14% of these loaves are thrown away as well. As you can see, bread is not only an energy-intensive, heavily processed item, it is extremely wasteful as a product sold on a mass scale for daily consumption. Significant waste is also associated with other grain-based products, such as pizza dough, bagels, muffins, waffles, and pastries, because the energy input and disposable waste generated are virtually the same as for bread.

As an aside, I used to work at a large grocery store chain when I was nineteen years old. Every day the bread men would come to survey the bread section for out-of-date loaves and literally dispose of garbage dumpsters full of bread products. These did not have to be molded or spoiled to be discarded but simply out of date. Much of the bread did not appear stale, and to keep people from picking it out of the garbage bins, fully wrapped and sealed, we were ordered to put Clorox bleach on the bread in the dumpster. Bread and pastry dumping is financially tolerable because large government grain subsidies make wheat cheap to grow and bread very cheap to make. It is not just bread we waste: most authorities believe we dispose of 30–40% of all food purchased, and what we purchase mostly is manufactured foods. Food has become as disposable as any other widget made by man.

In short, you don't find loaves of bread growing on vines or trees. Far from it. Bread production uses a large amount of Industrial Order inputs — in the form of fossil fuels, factory construction materials, and machinery to sustain the whole manufacturing and transport system — and this is nothing short of astonishing considering many people believe it is natural. Although this news is difficult for many of my patients to accept, the shared qualities do not end with the manufacturing process. Bread also shares the dubious distinction of cellular corruption that the other manufactured foods hold. As I pointed out in "Cellular Corruption," the consumption of grains (wheat, corn and rice) as they are incorporated into manufactured foods contributes to a significant amount of metabolic disease in the general population. Grains made into manufactured products are carbohydrates, and in effect are worse than most candy from a health point of view, being able to elevate insulin worse than table sugar does.

Bread products and other manufactured foods are most often calorie-dense, containing far more calories per gram than natural foods, and they also stimulate the brain reward circuits so that their consumption is reinforced. The raw materials and macronutrients that make manufactured foods calorie-dense, namely refined sugars and sometimes fat and salt, are the constituents that "reward" consumption by affecting the dopamine pleasure centers in our brains and as a result are considered "comfort foods," widely overeaten as a means to induce comfort feelings. This is obviously an artificial form of comfort since there should be little comfort gleaned from eating foods that corrupt our bodies while also prompting us to eat more.

Habitual unhealthy eating and emotional eating that adjusts moods has proven to create great profits for the food industry, but detracts heavily from the health of our population. After interacting with thousands of patients over several years, reviewing food diaries and analyzing successful weight-loss regimens, I realized that bread hurts more people than it helps. It is extremely common to hear a patient say, "Sorry, doctor, but I cannot give up bread." People are more likely to give up cakes and donuts than bread, even though bread and other grain-based flour constructs are nearly identical.

A simple rule I tell my patients that often has great impact is: No one should routinely eat foods that they have ever over-eaten! This rule appears to be rather simple and intuitively sound.

Sometimes we can find confirmation of what we believe in unusual places. I have seen hundreds of patients follow a low-carbohydrate diet and achieve substantial weight reduction and health benefits. Upon review of their food diaries (because I require this), I find the real reason for their weight loss was not the reduction of carbohydrates per se, but the reduction of processed carbohydrates and manufactured foods. My patients stated they were eating a low carbohydrate diet, and that may have been true, but what was obviously more important was that they were eating a low-bread, low-pastry, low-Frito, low-cake, low-Coca-Cola, and low-pasta diet. The diets were not described as being a "low-manufactured-food diet," but this is the reality.

Physiologically, there is absolutely no health requirement for grain-based foods. What is also very important is that consuming grain-based products means we are excluding more nutritious forms of carbohydrates such as fresh fruits and vegetables. Knowing what we know, it is difficult to believe that the food made in factories is really meant to enhance one's life, as in one's health. Therefore, when patients routinely ask me what to eat, my response could not be more simple. Don't eat manufactured foods from factories when we can eat whole foods from the ground. Eat what nature provides for us whole and don't tinker with the parts, calling it food.

To sustain optimal fitness in life, one must avoid manufactured foods because they are more likely to propagate robust revenue streams for the companies making them rather than propagate a healthy life. On the other hand, natural foods in their raw state, not modified by the Industrial Order, contain intrinsic value, meaning the innate nutritional value to propagate life. Naturally-occurring, whole carbohydrates (eaten in the forms in which they are grown) are not associated with our national epidemic of chronic diseases and multi-organ failure.

Fresh fruits and vegetables are real foods and do not instigate the feeling of comfort by way of neurotransmitter effects as grain-based products do. In fact, not one patient has come in my office and told me that she was grateful to be off the "fresh fruit habit", as if it were a comfort food (even though it should be comforting to eat). I did not see evidence of these patients binge eating apples, grapefruit, squash, and other whole fruits and vegetables. Eating fresh fruits and vegetables promotes weight loss as will be described in more detail later in this chapter.

I choose not to eat bread and other manufactured foods because there are much better carbohydrate sources for my health (fruits, beans, and vegetables) that are less calorie-dense, contain more water and fiber, and are therefore much more filling to my stomach. Fruits, beans and vegetables have more vitamins and antioxidants and no artificial preservatives. Foods from nature are the most authentic and sincere forms of nourishment, those being important distinctions to secure optimal health. Protein sources are important and will be described later as the main focus here is the distinctions of manufactured foods versus natural and the sources of carbohydrates I choose.

However, a warning is in order: while adding healthy foods to the average American diet may sprinkle some goodness amongst the toxins, it is insufficient action to prevent the overwhelming tissue corrosion associated with eating manufactured foods in general. We cannot undo the harms of manufactured foods by simply eating more healthy food. Adding natural foods to secure a goal of 5-9 servings per day, instead of substituting the good for the bad, in fact only promotes more eating. This indeed makes the government recommendation seem silly. Why not always eat fresh fruits and vegetables (perhaps this will be an impetus to subsidize the right food sources rather than the wrong ones), those items being our only source of carbohydrates rather than grains? My switch to a Naturvore identity promotes and sustains the best health nature intends for me because I am closest to my biological identity, the most sincere form of life. My Naturvore identity is my refuge from an industrial identity that will always await me.

NATURVORE POWER

The crux of healthy eating is to determine what is real food and what is artificial food (that made by mankind). From that point forward, our ability to understand food and its relationship to our health is greatly enhanced. I now choose my food without the drudgery of reading labels and contemplating the meaning of particular words. Instead, I can make decisions from across the grocery store. I can see what is artificial and what is natural from a distance. In effect, decisions of my health, such as eating, have become unintellectual and simplified a thousand-fold. My choices are not dependent on my intellect but on my values.

People will differ in their food preferences, which are both cultural and personal, but one thing is certain for all of us: we should love our food, not *just* for its ability to arouse our taste buds, but because it creates value in our bodies, hence making us more valuable to ourselves and society. Our choice of foods establishes and reflects our relationship with ourselves and our environment.

As we align our actions and therefore our lives to nature, we are more resilient by design. For me, Naturvore power (creating a healthy life, one biological moment at a time) is the most rewarding form of self-expression because I can create a more natural identity and a healthier life synergistically. The Naturvore identity represents the Natural Order within me versus an industrial identity wherein we have the Industrial Order within us.

Once I established a few definitions and principles to govern my interaction with the environment, eating for my health became more intuitive and second nature (or is it first nature?). As I have stated, we need not be trapped into the minutiae of food labels and micronutrients for a healthy life. Simplicity in our approach makes sustainability and consistency possible.

The most important dietary point I want to make in this book
is to avoid manufactured foods.

The Naturvore identity enhances health in the following ways:

It ...

- ... balances carbohydrates, protein, and fats from natural sources;
- ... infuses the power of nature into our tissues by consuming the building blocks of life;
- ... organizes personal biology and energy harmoniously to nature;
- ... promotes the propagation of life versus the propagation of industries;
- ... enhances our genetic heritage and the internal milieu of our cells, thereby creating beneficial attributes within our body;
- ... is anti-inflammatory;
- ... is high in anti-oxidants;
- ... is high in fiber;
- ... is high in micronutrients and vitamins;
- ... enhances gut flora;
- ... enhances immunity;
- ... prevents chronic diseases and cancer;
- ... avoids gluten;
- ... avoids glycemic problems;
- ... avoids food addictions and emotional eating; and
- ... avoids preservatives, artificial colors and artificial flavors.

THE BOTTOM LINE:

There are three macronutrients: protein, carbohydrates and fats; they fuel our activities and provide the building blocks for an optimal life. I eat them in a balanced fashion every day: generally eating 1/3 of my calories from protein, 1/3 from plant-based carbohydrates and 1/3 from healthy fats. I do not necessarily intend to eat fat with each meal as I will explain. Most people eat too many carbohydrates and fats with each meal, especially breakfast and snacks, and therefore we should start the coordination of our meal with protein considerations, making sure that we are in effect securing enough for our needs. Breakfast seems to be difficult with regard to sufficient protein for the average person. My breakfast is usually cottage cheese, ham or egg whites followed by a serving of fruit.

When considering each meal, I begin with a protein serving (4 oz.) and add 1–2 carbohydrate sources (both of these will be described in more detail below). Depending on the size and activity level of the person, eating 5 to 6 small meals a day (roughly 300 to 400 calories each) is more metabolically advantageous than 3 large meals. I have 5 and sometimes 6 small meals throughout the day, depending on how much I have exercised. I will begin my reporting with vitamins and then tackle each macronutrient separately.

Vitamins are secured mainly in my vegetable consumption, contained within a large salad per day with 10–20 constituents, as I am passionate about salads. Vegetables, such as lettuces, tomatoes, broccoli, cauliflower, are eaten mainly for vitamins, phytonutrients and fiber. Unless we place a bunch of croutons and other candied items on our salads (I don't recommend this), they really provide us more fat (from the dressings) and fiber than usable carbohydrates. In other words, one need not be very precise in considering the carbohydrate calories from vegetables like broccoli, lettuces, tomatoes and such because, when eaten whole, we are not likely to become obese on them.

The foods that appear most vibrant are the ones that provide serious vitality, and so I eat them every day. My salads are extravagant mixtures of ingredients sliced by a food processor (saves fingers) containing rich colors, such as the deep greens of broccoli, the reds of tomatoes, chard, radishes and peppers, the crisp white of bok choy, the creamy yellow of cauliflower and yellow bell peppers, the purple and white ribbons of red onions, placed over green beds of arugula, lettuces and/or spinach. Fresh beets provide a maroon tone to the whole salad but especially the juice at the bottom of the bowl. Jalapeño and habanero peppers lighten up my mood and provide additional health benefits, especially immunity. Portobello mushrooms add a basic taste, balancing the bitterness of raw broccoli and leeks and blending well with the rest of the ingredients. The whole dish radiates biologic goodness. Just thinking about the abundance of potent nutrients waiting to coarse through my arteries and veins provides a sense of optimism to enhance the marvelous taste. I want to feel and be as vibrant as the ingredients in my giant salads. I cannot cover all the excellent nutritional qualities within each of the individual constituents in my salads, but there is one, not frequently mentioned in the press, that I can use as an example to stimulate the interest of the reader to seek more information about what they eat.

My salads contain abundant mushrooms, and they are excellent nutrient sources. The amino acid ergothioneine found in mushrooms cannot be made by higher plants and animals, and so it must be consumed. Mushrooms are thought to be potent protectors against chronic diseases and cancer because of their content of beta-glucans, ergothioneine, selenium, and vitamin D. Beta-glucans have been shown to stimulate the immune cells by way of a cellular complex formation. Being water-soluble and able to cross the mitochondrial membranes, ergothioneine is very beneficial in reducing oxidative stresses related to normal metabolism, especially those that accompany intense workouts. Metabolism of glucose and fats is more efficient with abundant ergothioneine and recovery is improved because this powerful molecule also helps maintain the beneficial effects of other anti-oxidants, such as vitamins C and E. Mushrooms also contain substantial amounts of B vitamins such as niacin, pantothenic acid, and riboflavin that aid in the maintenance of our blood cells and blood vessels. Although a fungus, these edible delights sautéed in olive oil and sprinkled with garlic and pepper are quite delicious, simple and filling. I mostly consume them raw in my salad.

Foods that I would not have enjoyed in the past I now crave regularly. Deep-seated cravings now routinely occur not only for certain fruits and salads, but for Indian food (which is generally healthy and fresh) with heavy doses of turmeric, garlic, and peppers as well. Onions and garlic make food an aromatic treat but also impart cancer protection through the sulphur-allyl groups associated with them. Although most of my vitamins are delivered in salads, grazing on mushrooms, green beans, okra, spinach, and broccoli throughout the day can satiate a desire to nibble on something, to fill my stomach when in need of a snack. These are what I call "freebies" because they do not contain many absorbable calories, being ripe with fiber, vitamins and antioxidants, therefore are good between meals.

Protein requirements for optimal muscle function, immunity and health in general are somewhere between 0.6 to 1.7 grams of protein per kilogram of one's weight per day. I generally consume 100-140 grams of protein daily, but divided 5 meals. My protein sources generally come from turkey, chicken, fish, bison, beef, egg whites, and cottage cheese. Other sources of protein include lentils, black, kidney and pinto beans, and some people may rely on these more. I urge my readers to find what is right for them. However, the legumes tend to have more carbohydrates than protein and therefore are best supplemented with another source of protein.

It may not be fashionable, but I believe it is healthy to eat lean meat in moderation, just enough to supply our protein requirements. As descendants of the hunter-gatherers, we are omnivores. Therefore, it is biologically beneficial for us to eat proteins that resemble the fabric of our muscles. Combining extraordinarily potent antioxidants such as garlic, turmeric, black pepper, onions, oregano, and olive oil with meats is delicious as well as beneficial to my health.

The key to meat consumption is to eat only the leanest cuts and to eat only the amount one requires. I only eat red meat (beef) once or twice a week. Of course, the treatment of the animals — how they are raised and what they are fed — is an important considerations in my choice of meats. My recommendations for beef are that the cows should come from a farm that raises its cows on a pasture and feeds the cattle their natural diet of grass, from start to finish (most cattle in the US are finished primarily on corn, soy and other things to increase their density). Cattle consume grass as part of their biological heritage, and their evolutionary success is a function of that diet, although American industrial farms have hijacked the ideal as corn-fed beef. Entire books and papers have been written on the industrial corruption of the meat we get from cattle, and my first impression upon reading such material is that what has happened to domestic cattle certainly resembles what has happened to us. Human food, and our overall quality of life in biological terms, has been as corrupted as livestock feed and living conditions.

Pasture-raised cows have leaner, healthier, more natural bodies. Their meat is naturally lean and higher in protein, omega-3 fatty acids, vitamins E and C, and beta-carotenes. Their meat is also lower in saturated fat content than that of feedlot beef. Fortunately for my family, pasture-raised and finished beef can be found at the Hickory Nut Gap Farm, located in Gerten, North Carolina, close to our home. This is a small local farm where livestock is kept more compassionately and fed what nature intended, without artificial hormones and other industrial-scale gimmicks. I believe eating local food is a good practice all around.

Some have suggested that eating red meat is harmful not only to our bodies but to the planet, but this is more a function of the over-consumption of meat from unnatural sources than consumption of meat altogether. Eating large meat portions high in saturated fats combined with large quantities of processed carbohydrates is a well-described biological disaster, as discussed in "Cellular Corruption." Reconstructed meats, which include processed meats with excessive preservatives such as salt, nitrites, nitrates, and phosphates, are to be avoided since they are not healthy and hardly natural (assembled in factories).

Although there is scientific evidence that red meat consumption is harmful, no one has studied large numbers of people who have absolutely given up manufactured foods and eat strictly raw fruits and vegetables with their poultry, fish, pork or grass-fed beef and bison. In other words, scientists have yet to study thousands of Naturvores, and so I am quite comfortable eating protein from various meats that is not only identical to my proteins, but part of our evolutionary diet before Neolithic times. Eating adequate amounts of protein, especially in the form of meat, tends to reduce the insulin spikes associated with the carbohydrates we eat, insulin spikes being a cause of cognitive decline with age. Beef also provides an adequate supply of Alpha-lipoic acid, a water soluble amino acid that is shown to prevent cognitive decline and is also very important to our overall metabolic function, especially with regard to burning fat to usable energy.

Obviously, eating the appropriate amount of meat for optimal survival has very different health and ecological implications than eating too much meat. Indeed, most Americans consume too much meat for their own good and discard too much of the meat that is produced. American livestock is bred and processed as production units on an assembly line; these seemingly endless quantities make meat cheaper and ensure a perpetual production of chickens, pigs, and cows. The over-consumption of meat has nothing to do with the population's physiologic needs for a protein source, and therefore the over-consumption of meat products impacts the health of these animals (whose lives can barely be called living when raised on factory farms), the health of humans, and the health of the planet.

Meats also provide me flexibility and portability, in that I can easily take meat servings with me to fuel a high-intensity physical performance on outings. Turkey breasts or whole chickens are easy to roast and nice to have on hand in the refrigerator to be used when I need a serving of a lean protein. This way I can pack up my 4 ounce serving to take with me when needed. Estimating portion size does not need to be tedious and is easily mastered with visual cues. Four ounces of meat is roughly the size of a one-inch thick deck of cards or the area of the palm of a normal sized hand. However, I typically weigh my meats because I do not wish to short-change my body with too little nourishment. I have certain protein needs for optimal performance. The concentration of protein per gram is consistent between beef, poultry, and pork. Fish have a little less density of protein, and therefore I eat 5 ounces of fish protein instead of the usual 4 ounces from other sources.

Egg whites are an excellent source of protein. They generally contain 5 grams of protein per egg (my goal being 20-30 grams of protein per meal) but are also possibly the most efficient protein source to convert to muscle tissue. Proper brain function, especially a happy mood, is dependent upon the level of serotonin; and the high levels of tryptophan, an amino acid, in egg whites can be efficiently converted to serotonin (during medical school, I ate eggs at night to help with insomnia).

Cheese is a form of protein that can occur naturally, but most cheeses have a great deal of saturated fat and are therefore not a usual source of calories for me. I find low-fat cottage cheese with some fruit to be an excellently balanced breakfast on the run. Beans are a good source of protein but also a source of complex carbohydrates and fiber. They are very nutritious, inexpensive and versatile. Black beans prepared Cuban style are my favorite and I have them at least once a week. I will often combine egg whites with beans (which contain protein but must be supplemented for my needs) for a quick complete meal containing sufficient amounts of protein and carbohydrates.

When most people think of **carbohydrates** they think of the grain-based foods, but I think of fruits, such as apples, grapefruits, kiwi, oranges, pears, peaches, mangoes, grapes and berries. Fruits, butternut squash and sweet potatoes provide me *usable* carbohydrate energy for performance. I recommend avoiding the usual food fare of pasta, bread, crackers and candies (processed sweets) because they are manufactured. More nutritious forms of carbohydrates are readily available and are the plant-based foods such as fruits, vegetables and beans. Plant-based foods have been shown to be a staple in the diets of all the populations with highest longevity.

I eat about 1-2 servings of fruits combined with a 4 ounce portion of protein such as a lean meat in every meal except at dinner when I eat a large salad instead of fruit. A wonderful attribute of most fruit that makes life simple is that they all have roughly the same calories per fruit, size differences notwithstanding. For grapes and berries, I use one cup as an equivalent to a single fruit serving. Pure, unadulterated jams, perhaps crushed fruits, are also acceptable to the Naturvore as a carbohydrate source, but they should not contain man-made chemicals and preservatives. They are not preferred over whole fruits because they tend to have added sugars.

Sweet potatoes, and butternut squash have substantial carbohydrates and nutrients that are easily absorbed and can also be used in place of fruit servings for a source of essential carbohydrates, but the size of the serving should be reduced as one sweet potato is roughly 2 fruits servings in available carbohydrates. I do not eat these with butter, instead enjoying them microwaved as they are or sautéed in some olive oil. Before or after workouts I enjoy sweet potatoes heated in a skillet, as they contain a whopping antioxidant load as well as available carbohydrates.

The glycemic index is something that most diets talk about these days with the goal of avoiding sugar spikes in our blood stream, which induce insulin spikes and the metabolic chaos I have described in "Cellular Corruption." By consuming whole produce together with low-fat meats, the glycemic index becomes a tedious minutia because the proteins and fats I eat automatically reduce the glycemic load. Avoiding manufactured and highly processed foods does the same. Most cultures have a tradition of dessert after a meal, and so it is rather intuitive to eat your sweet foods like fruit after the more basic-tasting meats. Quite possibly this practice was born of better biological responses in our metabolism and digestion. It stands to reason that societal habits that are more or less consistent across the globe and across eons of time would have biological reasons.

Fats are essential to our body's functions, therefore we must eat them daily. Not long ago the term healthy fat was somewhat of an oxymoron, but our knowledge has changed. Monounsaturated fatty acids and polyunsaturated fatty acids (such as omega-3) are necessary for many biological functions (hormone production, for instance) and proper nutrient absorption. Maintaining a more balanced blood sugar by blunting the sugar-insulin spikes makes people less prone to inflammation, the metabolic syndrome, and diabetes. But fats are calorie dense, therefore we must use caution while grazing on them.

Fatty-acid molecules are also made into the major hormones, interleukins, prostaglandins, and other intercellular signaling molecules necessary for proper function of our blood vessels, immunological cells, and overall cellular integrity. Omega-3 fatty acids, especially eicosapentaenoic acid (EPA) and docosahexaenoic acid (DHA), are needed to improve the inflammatory response and reduce the chances for arthritis or other detrimental effects of inflammation in general. These fatty acids are my method of reducing inflammation after exercise instead of using non-steroidal agents such as ibuprofen, aspirin, and naproxen. Omega-3s have also been linked to improving mood and lowering blood pressure levels, and they are thought to decrease the risk of heart disease and irregular heartbeats. Sources for omega-3 acids include walnuts, but it is cold-water fish such as salmon (sockeye and coho), herring, sardines and tuna that contain plentiful DHA, which is an essential unsaturated fatty acid for optimal brain function.

DHA is important because it is good for the brain but must be generally consumed as the body does not make it readily from other fatty acid precursors. DHA supports nerve function, especially the white matter that contains axons wrapped in myelin sheaths. The nervous system is made of neurons and divided into white and gray matter, but the white matter contains myelin that allows for speedier transmission of impulses and more accurate information delivery. Therefore, children should have routine DHA sources. Omega-3 fatty acids are also found in other fish and seafood such as trout, anchovies, crab, and mollusks. One must limit the consumption of farm-raised salmon due to the PCB risks and also limit tuna and swordfish ingestion due to possible mercury content. It is a good idea to become familiar with species sustainability and the toxin risks of certain fish to make the wisest choices.

Other essential fatty acids such as linoleic acid tend to convert to omega-6 fatty acids (arachidonic acid products) and can promote unfavorable inflammation patterns. Egg yolks contain too much of this unfavorable fat. Free-range eggs may contain more omega-3 and so are possibly a better choice if one desires egg yolks. Unfortunately, a diet of manufactured and pre-packaged foods or fast foods will also provide far more omega-6 than omega-3 oils, especially if they are made with harmful vegetable oils, such as soybean oil. This a serious health threat in addition to the problems of sugar metabolism, insulin levels, and trans-fat loads associated with fast foods and manufactured foods.

Heart attacks and strokes are reduced in populations with higher omega-3 intake, which is thought to result from the reduced propensity for inflammation. Omega-3 molecules block the unwanted oxidation of cholesterol molecules, such as LDL, and beneficial cholesterol, the HDL form, is raised as triglycerides come down with omega-3 ingestion. HDL cholesterol increases with plant-based fatty acids, which helps to establish good vascular function by scavenging free radicals and neutralizing angry, oxidized forms of LDL.

My diet is so low in fat that I actually make a point to consume it. A tablespoon of organic almond butter or olive oil, a can of sardines or tuna can be eaten readily. A couple of tablespoons of plant-based oils are very beneficial on a salad because most of the antioxidants are fat-soluble and therefore will not be absorbed across our intestinal cells and reach the bloodstream or their intended destination without some fat mixed in with them. New salad recipes are a good reason to try new oils and salad dressing mixtures, and during the day I nibble on roasted almonds, Brazilian nuts, or walnuts to assure adequate fat consumption.

Following my Naturvore principles, I obtain nuts in whole or various natural or organic butter forms. These are squeezed into a paste or butter and not deemed manufactured because they are simply crushed, not reconstructed into a new product with additives. Look for raw nuts without added vegetable oils and salt and avoid adulterated nut butters made by factories that contain hydrogenated fats, various chemical preservatives, fillers, and even high-fructose corn syrup. Avoiding manufactured foods means we can avoid trans-fats, which are very toxic to our bodies. Remember, when eating whole foods we need not read labels.

We must not forget that the purpose of our nutrients is to fuel our essential biological activities, some of which are themselves necessary for proper nutrient incorporation into our bodies, perhaps as important as the nutrients we eat. We cannot expect to be well fed without incorporating good sleep, exercise and a healthy gut flora. Our relationship to nutrients does not end with their ingestion. Sleep and exercise govern important aspects of our metabolism and how our nutrients are used. Healthy gut bacteria (flora) functions as an organ in and of itself and is essential to the nutritive condition of a well-functioning human, and I encourage the reader to look into this a bit more.

Everyone should eat as much **fiber** as possible but a minimum of 30 grams per day. By securing our carbohydrates solely from whole fruits and vegetables, we are guaranteed sufficient fiber per day. I point out to my patients with chronic abdominal pains that they will feel better if they pay attention to what they are not doing, such as eating fiber and exercising, rather than attempting to micromanage their symptoms using pharmaceuticals. People with Irritable Bowel Syndrome (IBS) or diverticulitis have flare-ups due to the lack of a high-fiber diet and routine exercise (which tones the gut's nervous system and reduces stress) that causes these pains and disruption of normal bowel function. These conditions are another example of how the mistreatment of the body can induce both functional (IBS) and anatomical (diverticulosis) alterations in our organs. Many patients with IBS take pharmaceutical medications with significant risks and side-effects or will avoid certain foods, but a diet with fifty or more grams of fiber per day generally ameliorates IBS symptoms while also preventing colon cancer, diabetes, and heart disease. Another unfortunate issue is that many patients with IBS are misdiagnosed as having celiac disease or lactose intolerance because there is substantial overlap of the symptoms of these diseases. The importance of whole, natural foods with the nature-made delivery devices constructed of abundant fiber cannot be emphasized enough because our gut anatomy is such that it depends upon fiber for proper function and the appropriate timing of nutrient delivery. As with many aspects of our life, timing is everything

The contents of my meal fuel my performance, but timing is of great importance as well. I eat approximately every 4–6 hours and plan my life accordingly. I always have access to the time of day, and so I do not need to rely on my hunger signals to know when to eat. Proper eating is important to me, and therefore I do not skip meals. However, this calls for efficiency, and I find eating small meals to be efficient. Also, I must be somewhat flexible, therefore I generally eat within one or two hours of when I am supposed to eat.

Frequent meals allow for the hunger and satiety mechanisms to be in homeostasis. For most of us, modern-life demands can be rather unpredictable, but nutrient needs are very predictable. If I am more active, then I eat snacks between my meals, and if I am less active, I eat less. If I wait until I am really hungry, that means my body is likely breaking down some necessary tissues like muscle or even glycogen stores that I will need for the next workout. Eating frequent smaller meals is metabolically advantageous and prevents my body from feeling the need to conserve fat as if I were experiencing a famine. Indeed, I find I have much more control over my meals than I have over other aspects of my life, such as family matters and work requirements. A proper use of technology would be to assure the availability of healthy food, and there is abundant technology to pack meals effectively.

Nature has balanced the hormonal systems related to hunger and satiety for our advantage, and therefore we should use them to our advantage. At four o'clock, I may feel hungry, but if I know I will be eating again at five o'clock, then I can safely and confidently dismiss the feeling as not a true biological need. I can also pay attention to psychological impulses that may be misconstrued as a sensation of hunger. Do I have boredom or anger that needs to be resolved? Am I lonely and need to seek company? Do I need to do chores that I wish were not nagging me? Do I have irritation in my stomach because I ate jalapeño peppers in last night's salad? When we are bored, angry, tired, or feeling sad, we can feel hungry, and we are trained by the food industry to squelch these feelings with manufactured foods and the artificial comfort they provide. By divorcing myself from the most damaging effects of the Industrial Order, I began to see how even our most deep-seated realities can be altered by industries on a massive scale.

People frequently comment on their emotional eating and how stress can make them eat more, which at the very least indicates that hunger is a corrupted signal in America, one that has morphed into a feeling far removed from visceral realities and is now more aligned to our fluctuating moods. Authentic hunger signals were intended to signify the need to acquire and consume food before our bodies break down the tissue we have built. Many Americans seem to be hungry much of the time, but very few are threatened by emaciation, therefore we must question whether hunger is real, as in a genuine need for nutrients.

Emotional eating combined with the constant availability of snacks that manipulate our brain's cannabinoid receptors tends to make us a little numb and dumb. In fact, food may be contributing to the widespread apathy taking hold of our country, making us less inclined to rally our physical resources for exercise or to forfeit the relief from our feelings imposed by extrinsic pressures. I suspect, and much research has substantiated, that there is enough influence on our addiction centers in the brain to experience a food withdrawal by abruptly missing that bowl of chips as we enter the Mexican restaurant. Perhaps another snack cake, chips, or a bowl of ice cream will do to soothe our concerns for our future prospects. Americans report many reasons for eating poorly, but it is interesting to note that natural foods do not seem to fulfill those reasons. It seems logical to conclude that it must be something in the food.

The Naturvore identity is intuitive, and the principles are more important than the details. That is, if we eat by principles, we are more likely to recognize the simplicity of positive eating and more likely to derive value from our diet, as we would expect from nature — the details will take care of themselves. As noted, manufactured foods adversely affect our brains, and therefore we can't expect to think clearly if we eat them routinely, but I have found that life tends to take on a new perspective after opting out of industrial food madness, even when that decision is in its early stages. When we are more attuned to appropriate responses to feelings, more attuned to ourselves, self-care can actually become a priority. If we merely follow the principle of self-preservation, we will realize the immense power over our biology that humans have, a power to fulfill our needs rather than the needs of the Industrial Order.

As we understand the Natural Order better, we will know ourselves better. Eating brings us calories, but what is generally under-appreciated is the fact that the calories we consume are efficiently translated into information that is interpreted by our brains. Appetite and satiety are sensations produced through complex interactions of neurotransmitters, neuropeptides, metabolic factors, and psychology that wield profound influence over multiple organs, such as the brain, stomach, small intestine, fat cells, and the pancreas. The relationship of our bodies to our minds is reflected in our food intake, being very much tied to our values in life. When we eat, what we eat and how much we eat tells our bodies who we are and what our purpose is. I will only tease the reader with a brief explanation of the complexity inherent in neuro-hormonal satiety mechanisms. Perhaps a little knowledge of these intricacies could prompt an appreciation for and possible return to purposeful eating patterns. This is a window to how food affects our minds and our minds affect our food intake.

Under normal conditions (apart from eating a typical American diet or being obese) satiety hormones function very well and create the balance of true hunger and satiety — that is, until they are corrupted. Neurons have various receptors which are sensitive to the effect of these tiny molecules, therefore our digestive organs communicate directly (bi-directionally) to our brains and fashion our food intake to match our metabolic needs and our nutritional stores.

Bizarre sounding molecules like ghrelin, NPY, and the agouti-related peptide (AgRP) serve to stimulate appetite or initiate food intake. Ghrelin is produced by the stomach and the pancreas, and the level rises when we have been without food for several hours. It causes us to feel the pangs of hunger before our bodies engage in catabolism (tissue breakdown) in order to secure fuel for our essential activities. If we are calorie-deficient for long periods of time, then our cortisol (stress hormone) levels rise right along with our ghrelin levels. In such a state, we should expect to feel pretty bad and not likely to make the best choices with regard to food.

Our tissues, brain tissue in particular, will be crying out because we have simply neglected to feed them. Our resolutions to eat more healthily are abruptly broken when our hunger hormones have risen to such a degree that we must eat now. This explains why hunger is such a powerful drive that can sabotage the best intentions. The modern man need not live under the influence of ghrelin because we can plan our food intake if we wish. We will feel better and have a better relationship with ourselves if we keep ghrelin and cortisol levels down.

My feelings of hunger are easy to process when I eat by the clock and stay out of the ghrelin and cortisol effects. I get the common signs of hankerings, sour stomach or hunger pain, but I can assess if it is likely to be a biologically significant piece of information if I have sufficient knowledge of where I stand nutritionally. Eating on a schedule keeps the effects of ghrelin at bay, since I usually eat before I am hungry, and if I am feeling a bit of hunger, say an hour early, I am reassured by the nutritious meal waiting for me. Upon understanding how our bodies can be optimized when eating on a schedule, it seems that waiting for true hunger is self-defeating behavior.

Satiety can be secured more quickly with fewer calories consumed by eating natural, whole foods. Our digestive system evolved under the influence and direction of the innate properties of natural foods. Nerves in our stomachs and upper intestines activate with the stretching and distention of the stomach and signal the brain to stop eating. Fiber and other properties of foods from nature make them generally less calorie-dense, therefore our organs are more likely to signal we have indeed eaten. A series of peptides and hormones are also triggered by the digestive organs to provide our brains valuable information. Insulin, triggered by GLP1, is then secreted by the pancreas into the bloodstream in response to sugars and not only provides satiety signals in the brain but also promotes the entry of sugar molecules into our cells for fuel. The typical American diet corrupts the function of our hunger-satiety mechanisms, making our organs less responsive to the intracellular signaling of insulin, leptin and other hormones like cholecystokinin (CCK), those intended to give us the sensation of satiety after a meal. Eating manufactured foods tends to make us less content, even after eating more calories.

Leptin is a hormone that has received much attention from the scientific community because it promotes the sensation of satiety and is also nature's way of stimulating fat loss. It is produced by fat cells proportional to the amount of fat stored, therefore as our fat accumulates, the levels of leptin rise; the purpose is ultimately to suppress our appetite (a natural appetite suppressant). Leptin is a hormone that actually promotes a lean body. But as we should expect, the information that leptin is supposed to be giving us is lost as we pursue other reasons to eat than nourishing our lives. As one gains weight, leptin levels rise and then remain high, which causes receptors, such as those in the brain (hypothalamus), to become less sensitive to the effects of leptin and therefore resistant to its effects. Unfortunately, leptin resistance means that one doesn't feel satiated when one should, and so one cannot feel full after a large meal. This is the same problem that occurs with regard to insulin, which also remains elevated, and the insulin receptors become desensitized to its effects. Everything tends to be in good working order as described until we become unnatural by eating processed foods and gaining excess weight. The typical American diet of manufactured foods corrupts these elegant biochemical systems from their original purpose and suffering ensues. Excess circulating insulin promotes inflammation, cholesterol deposition, cancer growth, immune dysfunction, diabetes type 2 and fat deposition. High leptin levels are also being studied as a cause of inflammatory conditions such as arthritis.

Fortunately, insulin- and leptin-sensitivity can be reestablished by losing weight, exercising, and eating a diet of natural whole foods. That is, one can improve the sensitivity of one's receptors to insulin and leptin by improving one's diet and exercising more to reduce obesity. Satiety/gut physiology tends to remain in order by eating small meals that are high-fiber, low-fat, low-glycemic, and therefore not calorie-dense. However, as I suggested earlier, this is not as complicated as it sounds. All we have to do is avoid manufactured foods and we are sure to find optimum nutrition without much thought. We need not be intellectual about eating if we remain sensitive to the information provided by our most ancient body systems. Being receptive to the goods of the Natural Order brings life around full circle.

As a Naturvore, I eat only whole foods such as vegetables, fruits, and lean meats that pack high nutrient, water, and fiber content. By so doing, my normal satiety mechanisms are preserved and I feel appropriately full after I have consumed only 400 calories as opposed to a normal American meal of 800–1500 calories. Viewed the other way around, it is extremely unlikely that I will overindulge on raw fruits and vegetables because they are full of fiber and water that cause gastric distention.

By eating natural foods and eating when we are supposed to, we can manage our digestive hormones to serve us well, performing as they were intended. Calorie-dense foods, especially manufactured foods, diminish our cellular health and prevent satiety before large quantities of calories are consumed. This is a consistent attribute of artificial foods, and their abundance corrupts the bodies of hundreds of millions of people. By contrast, natural foods do not alter brain chemistry in such a way that induces people to continue eating them for the sole reason of mental comfort.

Eating for nutrition is part of my well-formed identity, and it is the best guarantee out there to be resilient to disease. Proper, vital nutrition is quite exciting and therefore I have a renewed passion about eating in general, but for completely different reasons than before. Securing vital nutrition is a vital purpose for my life.

Vital foods give rise to vital people.

I hasten to note one more time that the Naturvore identity, while utterly intuitive and therefore easy to maintain, is not a hit-or-miss affair. So-called "healthy" foods in isolation are like diet fads geared to treat symptoms and syndromes rather than promote general well-being. Nevertheless, we are consistently exposed to messages about a given food as "heart healthy" or "good for brain health" or that it "helps arthritis," and other such narrow-minded suggestions. The truth is that any food that is truly good for heart health is also good for brain health, liver health, and joint health. What is good for one organ system should also be good for the others, or we should not be eating that particular nutrient at all. Such claims also defy reason, for if a heart-healthy diet is good after a heart attack, why not serve it before the heart attack?

The most time-consuming and complicated dish I prepare routinely is my salad, and yet, with one simple dish, I can prevent most cancers, heart disease, strokes, high blood pressure, arthritis, diabetes, and Alzheimer's dementia at the same time. No thought is required, or complicated math, or chemical analysis of micronutrients. I simply mix green, red, white, yellow, orange, purple and gold colored vegetables, mushrooms and tomatoes (sometimes other fruits as well), mushrooms (fungus) and vegetables in a bowl with ground ginger, some olive oil, a dash of balsamic vinegar, and perhaps other titillating spices, such as habanero peppers, sprinkled on top. The time it takes to make a salad is worth it to me, a routine for health as if I were prescribed it. There is a certain intuitive satisfaction to eating different classes of fruits and vegetables that approaches wisdom. The colors themselves symbolize the different nutritive effects, and my performance serves as proof of those effects.

It does not require intellect to know the difference between the various effects of the Industrial Order food offerings and those of the Natural Order, and in fact it is our intellect that might convince us otherwise. To put it simply, it could be said that I have a balanced diet because I balance the colors and textures of fruits and vegetables, vary my source of proteins and fats, and eat various condiments. A statement of dietary choice does not get much simpler than that.

Again, however, the key is to adhere to the principles underlying my Naturvore identity. Within my most natural identity, the ethic of self-preservation resides within me at all times, and thus I am at much less risk to do myself harm by eating poorly because I am intent on being healthy. For instance, when offered manufactured food at social gatherings, I will say, "No, thank you — I eat only natural foods." I feel much more comfortable in my own skin when I walk away from a party or social gathering knowing that I am still biochemically at an optimal level and ready to perform whatever physical activity I may wish to accomplish. The ability to live a successful, healthy life is an asset that can be acquired through intention, but this intention is self-reinforcing, for I enjoy the mental and physical benefits of my good health, and this has changed the entire context of my life. This is the Naturvore identity as it is lived, and these are the basic tenets of the Naturvore identity simply stated: we individually have power over our own health, and this begins with the recognition that the food we place in our mouths should definitely not corrupt our form, function, or biological purpose and thereby make us people we do not wish to be.

SELF-DISCOVERY

As I have found out, our health and our physical form and function are the ultimate strategies of self-expression. I suspect most people do not consider their physical form to be a choice, but the proof is all around us. We choose what we put into our mouths and we choose the level of exercise we will perform, therefore we not only choose our physical form but our health as well. In fact, we have a greater range of options as regards our physical form than any culture in history. As a nation, we tend to celebrate our liberties and the idea of abundant choices, but when we consume the products of the Industrial Order, the creator of so many choices, we indeed assume particular conditions and forms as well.

American technology, as delivered through manufactured foods and in the form of life-support techniques, such as home health services and sporadic medical rescues, allows the range of physical size among this nation's population to be quite extreme. In fact, America is home to the world's largest population of really large people. In the hospital, it is routine to see people who weigh more than 300 pounds and cannot fit into CT scanners and MRI machines. It is not uncommon to evaluate and treat people over 400 pounds, and there are even patients who weigh 500–700 pounds. When a person's size is so distorted, the context of his life and the context of my medical care change. For instance, I am often at a loss for words when such patients ask me what is wrong with them, no matter the complaint.

Already, more than one American out of twenty carries more than 100 pounds of excess fat on her skeleton; this population has more than doubled since year 2000 and is still escalating in prevalence. New medical concepts and terms are being engineered and applied to these patients, such as the "super-morbidly obese" (a person with a body-mass-index greater than 50 — less than 24 is normal). Within the range of physical size options in America, the sky is the limit, and to hear the way some people joke about their size leads me to believe that a person's size is indeed a form of self-expression. Through the synergism and symbiosis of manufactured foods and manufactured medical devices, virtually any American can secure any size or fitness style they want delivered with factory precision. For instance, I have proof that I could be larger in size or even more atrophied, if I wish.

Scientists frequently discuss our genetic differences with regard to our physical attributes, but that completely ignores the unified propensity of 300 million genetically diverse people to be rounder forms rather than more linear forms. Indeed, although we appear to have a great range of options in our physical form and performance, we are pressured as a herd to become a uniformly rounder and heavier version of ourselves. Leaner Americans, who even so carry roughly 25–40% fat content in their bodies, are still composed of corrupted tissues because of their diets. Our choices do not end with food. A visit to the doctor will provide us more choices through access to a large selection of methods intended to artificially tweak (through prescriptions and procedures) the sinister disruptions of our biochemistry quite readily, but these interventions will not fix the underlying problem — that is, how we treat ourselves. We tend to compound our bad choices.

In America, we can change our physical forms throughout our lives, and many Americans experience this by virtue of weight cycling, the abrupt gaining and losing of significant amounts of body-fat weight. Changes made in our energy balance directives that are either not sustainable or undertaken for temporary reasons, such as a high school reunion, a wedding, summer vacations and such, instill a less than authentic form of existence to our corporeal condition. Perhaps this is simply another way that American ingenuity (for example, the many unsustainable diet plans created each year) creates artificial conditions for us to buy into. Technologically-based diets (fad diets and calorie gimmicks) simply provide the means to enhance our personal range of options over our physical forms, but only on a temporary basis. In fact, researchers have concluded that weight-cycling behavior creates worse health conditions than remaining obese. I suspect that whatever harm can be confirmed by blood analysis or other testing is only the tip of the iceberg and that the real harm is done with regard to how these persons live their own lives. Furthermore, weight-cycling, ingestion of diet pills, undergoing procedures and incorporating other tricks of the Industrial Order are not likely to enhance how we live with ourselves or represent the most authentic life.

We can only realize our true identity when we secure it naturally.

Anyone can be as confident in his choice as I am confident in my Naturvore identity, and therefore he or she too can opt out of the herd effect and the promises of the good life through abundant technologies. Despite the pressures from the herd to conform to a less than ideal physical existence, we can choose to be different. We can choose not to ignore the greatness of the biology, the natural aspects of our lives, that we all possess.

SURVIVAL AND THE FITTEST IDENTITY

We tend to think of our identities as a macroscopic form, largely a presence created by the Industrial Order, but what is more true, and therefore more important, is that we have a microscopic identity first. Furthermore, our fitness is an important aspect of our identity and is dependent upon the health of our tissues and upon our capacity for purposeful energy use, a topic that I will visit in detail later. In nature, an organism's form efficiently dictates its functional capacity, its overall fitness and its resistance to disease, while the converse is also as true, that function builds the form. In fact, it is safe to say that physical form and function are inextricably linked in nature. Hence, humans who carry excess calories stored as fat and are deviating from their more authentic human form are also distorting their function. Expectedly, their health tends to deteriorate and their personal experiences in life change.

The simple balance of input and output with respect to calories governs much of our survival fitness. There is an obvious benefit to having some nutrition stores, but that benefit is readily negated by excess stores. Total calorie consumption is important in our lives because it is very clear that, the more calories we ingest beyond what we need for optimal survival, the greater the loss in health and diminishment of our lifespan. It is well established that reducing the caloric intake of any organism by 25–30% — and this is true all the way down the evolutionary chain of complexity to the fungus — will increase longevity. This is the crux of the calorie restrictive (CR) diet and lifestyle, a proven way to a longer life with less disability.

The term "calorie restrictive diet" carries negative connotations perhaps, and so instead maybe we should consider such a diet regimen, one sufficient as regards caloric intake for our lifestyle, "calorie appropriate." However, to not indulge in excess calories on a daily basis is not only unusual but (dare I say it?) un-American and possibly exhibiting a hint of an antisocial predisposition. Over-indulgence is, after all, seemingly as American as apple pie and baseball.

In fact, I would argue for the term "calorie appropriate" from my own case. I am more active than the average calorie-restricted person, and my body supports more muscle than I have seen on people following the CR diet. Therefore, the calories I need to consume to maintain the physical structure that I desire are more than the calories consumed by other restricted people. However, although I probably eat more calories than most people on a true "calorie restricted" diet, I still maintain approximately 10% body fat, the lean physique associated with calorie restriction.

My diet is appropriate for me because I eat enough food and nutrients for exceptional performance while maintaining the lean characteristics that assure longevity and good health. Therefore, for me a well-defined physique is an essential measure of how well I am matching the calories I take in to my activities. The lean state is always by definition a calorie-restricted state, because there are no excess stores. Famine has historically been one of the gravest threats to our survival, but now it is food manufacturing and the unprecedented quantity of convenient, calorie-dense food that threaten our lives. From an evolutionary point of view, the satiety hormones within each of us have never had to respond to a manufactured food diet nor such a plethora of dense calorie sources, and long before our hormone systems could be expected to evolve in response to the current food situation, the human race will be extinct from such practices if we continue down this path.

Ancient humans, the people from whom we have descended genetically, had to perform much of their hardest work when they were very hungry and tired. Our ancestors had to run from danger and take down food even while they were enduring real starvation. Obviously, to survive, our biological systems developed mechanisms that provide us better performance even when our reserves are depleted. In the wild generally, in fact, when times are tough, an animal must perform better than normal to survive. This is seemingly counter-intuitive to how most of us live now. Our society tends to coddle people instead of promoting physical resiliency, and the results are obviously the pandemic lack of health we witness around us.

Calorie restriction (CR) has been thoroughly studied by scientists as a means to enhance or preserve our performance and longevity. CR works because it applies stress on the body that induces the transcription of genes especially for this state, genes that are present in most animals and preserved across most genus divisions of the universal family tree. Researchers have posited that genetic entities such as the DDR complex, the sirtuin genes, and the heat-shock genes provide the mechanisms so that we successfully adapt to stress. By virtue of an enhanced ability to handle physical stresses, we are conveyed longevity because we are working at peak efficiency.

With the exception of the nervous system (which tends to preserve cells and therefore is consistently involved in housekeeping and clean-up functions), the tissues of our body are constantly dying and being replaced. Efficiency in cell turnover and cellular clean-up is a large part of the body's health. As cells age, they tend to acquire cellular debris (worn-out organelles) and other dysfunctional characteristics, and therefore they become littered and more prone to errors and dysfunction. One way to fix cells that are littered is to implode them and digest them, meaning they are killed and their constituents recycled. Appropriate cell death takes care of this problem by not allowing cells to live long enough to acquire disastrous age-related changes that would adversely affect the organ's function. Timely cellular death that is part of our normal development is accomplished by the cell itself and these stress genes regulate this ability. Indeed, apoptosis (autonomous cell destruction) is quite advantageous because it renews our tissues and we become fresh again. But if we are imploding our own tissues as a means to renew, we had better be efficient at it, and CR has been shown to make apoptosis more efficient (during stress times, we need the best tissues). In effect, we can govern appropriate apoptosis by our choices in life. Calorie abundance tends to make our cellular mechanics sloppy and leads to errors in very important biological ways, affecting our performance and our longevity. Baby-fresh, newer cells tend to last longer and perform better.

In neurons cell turnover is less advantageous than in most other locations in the body. Instead of spending energy to orchestrate routine cell death, neurons engage in robust cytoplasmic repair work, and this is also enhanced by CR. In summary, restricting one's calorie intake improves quick cell turnover mechanics but also enhances intracellular clean-up activities in the cells that are destined to have lengthy lifespans, which means that our caloric intake tends to determine how well we live now (present performance) and our performance over time (longevity).

Self-care translates efficiently to better genetics and should be considered the ultimate form of self-expression by all.

The strong influence we have over our genetic pattern is how our personal responsibility is put into effect in our lives. We determine many aspects of our DNA transcription, the method of DNA-template reading that creates our proteins and hence our lives. The stress gene activations I have previously mentioned are just a few of the important mechanisms. Our actions, and even our thoughts, can determine which of our genes are transcribed, how they are transcribed (portions versus whole) and when they are transcribed. We can exert strong, life-changing influence over our natural fabric by altering the mechanics governing this biological template of our life. Self-care becomes self-discovery when it is done properly.

Strands of DNA resemble a ladder that is twisted by the electromagnetic forces (the propensity for attraction versus repulsion) that are present due to the polarity of molecules. The sides and rungs of the ladder are built by four base pairs: adenine, thymine, guanine, and cytosine molecules. The strands of DNA are then coiled around histones (proteins), then super-coiled into the shapes we call our chromosomes. From this basic template of a biological life our lives are sculpted. The DNA template may have been given to us, but our lives are indeed our own because our actions ultimately determine how the majority of the template will be read and utilized. In other words, the strands of DNA determine our potential, but we determine how much of that template or potential is realized. DNA reading is influenced by the American cultural landscape, but only if we choose to expose ourselves to it. If we "opt in" to the cultural madness, caloric excess combined with chronic sedation and/or a sedentary status, then we are exposing ourselves to the environmental influences that determine how our DNA template is read for the worse. We have vitality genes within our genome, but we must be vital, act vital, to access them. Indeed, our vitality genes only serve vital creatures.

The genome or DNA we are born with is not a stagnant, rigid digital code. Genes themselves have activation zones (promoter regions) that are influenced by micronutrients and activity demands. Therefore, the choices we make alter the frequency of particular genes being expressed. DNA transcription results in the formation of messenger RNA (ribonucleic acid), which leaves the nucleus of our cells and travels to the ribosomes, protein complexes that produce functional proteins to sustain the structure and mechanics of our lives. RNA codes built from the DNA codes specify the amino acid sequences that are put together to make the proteins. In summary, the DNA is translated into RNA and the RNA into proteins that serve structural functions, molecular transport functions, enzymatic functions, and even turn into the antibodies that fight infections.

Our lives are, generally speaking, very unlikely events at all. We are held together by these systems that fight the universal force of entropy, but within the scope of these biochemical reactions, the special events such as the trillions of enzymatic actions of proteins, our lives are translated from an instance of improbability to direction and meaning. How we treat our DNA, those cellular moments I mentioned previously, is how our lives turn out, at least more often than not. There is no way to escape taking responsibility for our lives because much of how our lives unfold is actually a result of how our DNA is unfolded.

Epigenetic is a term applied to the genetic control mechanisms that lie outside the genome, and these controls largely determine how much, how often, and how long DNA is uncoiled. Uncoiling of the genome allows the genetic promoters, gene regulation assemblies and transcription assemblies, to work on the DNA template. Epigenetic changes can have fantastically different results based upon how we interact with the environment, and thus more precise and immediate control over our genome and our lives can occur through epigenetic means. Factors in our lifestyle such as toxins, diet, exercise, and even our thoughts will alter the epigenetic components, usually appending informational tags to the strands of our DNA. We are made real beings from our minds to our toes by these DNA mechanics I have described. Because we tend to choose much of our fate physiologically, I find this quote from the twentieth-century Spanish philosopher José Ortega y Gasset (1883–1955) to be quite inspiring.

> We are not launched into existence like a shot from a gun, with its trajectory absolutely predetermined. The destiny under which we fall when we come into this world . . . consists in the exact contrary. Instead of imposing on us one trajectory, it imposes several, and consequently forces us to choose. . . . To live is to feel ourselves *fatally* obliged to exercise our *liberty,* to decide what we are going to be in this world. Not for a single moment is our activity of decision allowed to rest. Even when in desperation we abandon ourselves to whatever may happen, we have decided not to decide.

We are always making choices that affect our vitality because our most immediate actions result in DNA transcription by way of the epigenetic control mechanisms. For instance, choosing artificial food, with its calorie density and unnatural qualities, induces adverse DNA transcription, largely by these epigenetic means. Choosing to be sedentary also changes how the DNA in our cells, not just in the muscles but the liver and other organs as well, is read. We exert favorable influence over our DNA by exerting ourselves. Unfortunately, passive, physically unengaged patients often suffer more than physically active patients, even during medical treatments, because doctors do not wield the power over our DNA mechanics that we the patients do.

The DNA we possess can be transcribed to a person's advantage, producing wellness, or it can be transcribed disadvantageously, producing sickness. For example, I have developed a much improved physique and physical performance level over what I had a few years ago — or at any time other time in my life — not because my genome changed but because the way it is transcribed has been changed. The environment that I live in has not changed dramatically during the time that I have dramatically improved my performance and physique, but I changed my interactions with American culture. The better biological performance that I have now was inside of me as potential and was only realized as I put forth the effort to use it.

Perhaps a question we could all ask ourselves is: "Why should I let American culture bring out the worst genetic fate I have in store?"

Once the environmental effects that were corrupting and suppressing my authentic biological potential were removed, a more authentic me unfolded, literally. I liken it to a personal genetic evolution that occurred over months rather than the eons it takes for the species to evolve. Proof of such profound genetic alterations does not require sophisticated testing. Simply look at the log books of my bicep-curls performance or my running times and you can discern the reasons and the evidence that my physical function has changed, not to mention my physical form. But to satisfy readers in need of scientific confirmation, I will point them to an article in the journal *Cell Metabolism*, wherein researchers report, "Exercise elicits gene expression changes that trigger structural and metabolic adaptations in skeletal muscle" (*Cell Metabolism* 15, 405–411, March 7, 2012). Improved physical performance results in more performance- driven form, a result of improved DNA management.

We routinely express ourselves through epigenetic mechanisms.

Likewise, though conversely, obesity and a sedentary existence, practiced states as much as physical fitness is a result of fitness practice, change our DNA transcription and bring about a host of bad outcomes for life. The foods we eat and the toxins we expose ourselves to invariably alter us in profound ways, and because our minds are a confluence of energy transfers, they can be altered by the contents of the messages we absorb. The mind works as a result of synaptic transmissions. That is, these transmissions create our brain's milieu and therefore the contents of our minds.

Epigenetic mechanisms also work on our minds, as one would expect, and therefore the mind can be trained to believe or think in a manner that is very similar to how muscles adapt. The mind's transcription effects can often result in the most powerful alterations in our bodies because the brain controls the greatest flow of energy and information. When the mind is made up, things can happen very quickly because there is a focus of all other energies in the body. In effect, there is an alignment of our genetic information and we can use our biological heritage the best or worst way possible. In short, control of our thoughts means control of our DNA, our metabolic activities, our food ingestion, and so on throughout the body. Inherent to the human species is our ability to choose most aspects of our lives. "Life is what you make it" is indeed the truth, and at the most basic level of who we are.

The American cultural landscape is filled with legitimated toxins in the form of the foods we eat and the pharmaceutical products that mask the effects of those foods. These toxins act at the level of our DNA transcription. Manufactured foods are not sold to us as gadgets and fabricated products that act on our DNA, but that is precisely what they are and what they do, whether we choose to consider the ramifications or not. If we have doubts about my concerns, we can usually look into the mirror and see the fact that we have become human forms that we never intended to become, a real life manifestation of basic biology and DNA transcription gone awry. Or we have simply chosen the form and everything is according to our plan.

Even if one possesses cancer genes, not everyone gets cancer. Smokers and obese people are not born with cancer, but their actions contribute to an increased risk of cancer developing within their cells as a result of changes made to their DNA. Therefore, instead of thinking that our fates are precisely predetermined genetically, we really must keep in mind that much of our health fate in life has yet to be determined. Controlling our body weight and lifestyle habits can change our health fate quite favorably. Unfortunately, most people with cancer-associated genes, the familial cancers, tend to engage in the same lifestyle habits and have the same body shape as the cancer-afflicted people in their family.

Personally, I find it reassuring that my important systems for life, such as immunity, strength, agility, and metabolism, are maintained by genetic means but are sculpted by me. In fact, we can call upon these systems to make us more efficient and stronger, and this is the proposed mechanism for the beneficial effects of a calorie-appropriate diet. A calorie-appropriate diet enhances health and longevity by enhancing the mechanics of these genetic rescue systems. By contrast, excessive calories will not only cause adverse cellular milieus that result in conditions such as Alzheimer's, DM, obesity, and cancer but will also suppress the genes responsible for cellular rescue and defenses. A manufactured food diet, laden with processed sugars, has even been shown to suppress the p50 genetic mechanisms responsible for eradication of cancerous gene changes.

We build ourselves through physical stress by way of physical adaptation. With this in mind, I recommend that everyone, but women especially, should run, jump, and play vigorously to build stronger bones and prevent osteoporosis. Most people instead ingest nutrients, supplements, and even pharmaceutical medications, foolishly neglecting the most important inducer of genetic adaptation, physical stress on their bones. In fact, regular physical stress on our bones, in concert with consuming calcium and vitamin D, is the best way to prevent osteoporosis. Regular exercise causes the cells that model bones, the osteoblasts, to lay a sound foundation of mineral apatite, the cement. However, regular strenuous exercise will improve not just these adaptive changes in the bones but will also induce genetic enhancements all over the body. Again, thinking in a holistic fashion, the fix for osteoporosis tends to fix the other problems related to poor health, which is not accomplished by taking osteoporosis medications.

Obviously, most of us need to plan for physical stress because our lives do not include physical stress either in the workplace or at home any longer. Modern life tends to promote more mental than physical stress, and mental stress should be avoided because it breaks our positive spirit and distorts our reality all the way to distorting our physical being. In fact, there are a great many physical pains seen in the doctor's office that are associated with mental stress. Going through life with our fight-or-flight system activated causes us to age prematurely. If we go through life with negative thought content, then we have summoned adverse neurohormonal responses that tear us down and make adequate immunity and healing very difficult. Under mental stress, we tend to react negatively rather than remain in a positive flow, expecting the positive. Physical beings can expect a better life by being physical, not just by thinking or worrying about the "good life."

Fortunately, choosing to exercise induces a positive flow of energy, one that is known to unfold our DNA in a positive way, thereby placing us closer to our intended physical form. Although energy is required for our form to take shape, our form then governs our energy patterns. This reminds me of the chicken and the egg conundrum. Vigorous exercise strengthens our minds and therefore improves our view of reality because our circumstances are put in better perspective, beginning with the fact that our physical health has just been improved. When we act with vigor (exercise) we become more real beings by providing a vital purpose for our tissues. The power inherent in an authentic and positive life commands our tissues in such a way that we (our mass or physique) are optimally adjusted to transmit the force of life.

A LIFE FORCE

Every organism makes energy, the defining attribute between the living and dead. Our lives are simply patterns of energy and these patterns are dictated by our form and our function. Life forms have three very distinct characteristics that separate the animate world from the inanimate: self-assembly (able to repair and procreate), self-determination (able to direct energy) and self-awareness (able to distinguish the self from other things). This triad of characteristics is created from energy use, the capture of energy made possible through our physical forms which are dictated by the Natural Order. We all transmit the force of life, but some of us have more resistance — meaning we have created impediments to our current of life.

My original resistance to the Natural Order was a result of the Industrial Order in my life. Manufactured foods and the propensity toward a sedentary state corrupted my physical identity and therefore also my ability to use my life energy for health, since our tissues govern our energy profiles, as I will show. I coined the term "Naturvore," and consider it a new identity, because I could sense a stronger, more coherent life energy within me, a profound change from the biological chaos of earlier times. Power from the Natural Order places us in a very positive set of circumstances for our health and our happiness.

Within the biological world, an organism's activity (expenditure of energy) produces favorable outcomes more often than not. But this is not the only aspect of energy management attributed to people for there is another held in optimism itself and the alignment of one's life to others engaged in life-enhancing activities. Most people would not likely consider the power contained in positive life energy to be tangible, like the horsepower derived from the combustion of fossil fuels or the current of electricity coursing through our houses, but the favorable influence over one's circumstances derived from a positive action, mood, personality, and belief system is undeniable. I will show that positive power need not be so mysterious.

All life forms are able to mold the inanimate world to suit their needs. Even creatures that are microscopic arrange and rearrange dirt, rocks, air, water, and other physical objects. The entire surface of our planet and the atmosphere above it have been fashioned to the desires of life, the biomass being perhaps the strongest force for planetary change. That is, life does not sit idly by but exerts a force upon the world. The force of life is usually denoted as mystical, and yet we are made from it and can feel our own power and substance. The force of life has yet to be fully defined by mathematical equations measured in internationally standardized units, but its presence is undeniable.

Life requires energy to combat the universal force of entropy. In fact, the organization of elements and biological molecules into an organism requires energy input because the process defies entropy. Therefore, all organisms are specialists in energy management, not simply using just enough to organize their bodies but setting aside reserves to endure hardship. Organisms can also wield their energy to pierce new environments and succeed against all odds in a hostile world. Energy management is an important endeavor for the living because life is not simply a steady state or position of neutrality. Organisms are either on offense, on defense, or repairing themselves. We tend to live our lives oblivious to the realities of personal energy management, and we suffer as a result.

The human body derives its power from the molecular utilization of adenosine triphosphate (ATP). Virtually all of the cellular machinery on our planet is either directly or indirectly coupled to the energy captured from the breaking of phosphate bonds as ATP becomes adenosine diphosphate (ADP) or adenosine monophospate (AMP). The difference between these forms of adenosine molecules is the number of phosphate groups attached to it. The loss of one or two phosphate groups supplies the power to trigger the necessary reactions for life and the force of life acting on extrinsic things, such as other organisms, or to mold the surroundings to suit our needs. The conversion between the three forms of adenosine occurs instantaneously, and so their concentrations are in constant flux, a condition that has the properties of a power current.

The nutrients we consume are converted into transferrable power, the energy held in the ATP-phosphate bonds. Concentrations and ratios of ATP relative to other metabolites resulting from its breakdown, such as cyclic-AMP and ADP, determine the activation and speed of critical cellular events such as DNA transcription and RNA transcription. Other essential functions performed by ATP are also related to energy management, including free fatty-acid transport carriers, cellular signaling complexes in the cell membranes that are crucial to communication between organ systems, hormone responses, and all the other important cellular activities that are needed for a successful life. ATP is not stored but instead always being generated and used in response to need within a constantly changing cellular environment. Its presence is so fleeting and free-flowing that I like to think of it as having both effervescent qualities and current properties, omnipresent and universal but also always appearing and disappearing, constantly flowing, ebbing, and fluctuating to meet our needs.

The balance of ATP to ADP phosphate bonds provides the power for a sperm to whip its tail and reach an egg, for our brains to create thoughts and moods, and for our gut to absorb nutrients and our muscles to carry us through our daily activities. ATP combined to the enzyme luciferase provides illumination to organisms like the firefly so that it can dazzle its mate, and ATP provides the flagellar locomotion for bacteria. Even the readers' thoughtful reactions to my words are made possible through ATP-generated currents along neurons within their brains. Although there is little capacity to predict such a small event with such fleeting qualities, the effects may last one's lifetime and have far-reaching implications. ATP-generated information is quintessential power and beauty.

By virtue of the interactions between ATP molecules and the genes contained in DNA and RNA (the nucleic acid chain that transmits genetic information between DNA and the protein manufacturing sites), organic molecules become incorporated into organized patterns of tissues, organs, and finally organisms. ATP is also readily converted into the molecule adenine (by removing the ribose sugar), and this molecule builds DNA and RNA strands. Therefore, ATP transfers biological power by two methods, through the energy of a charged phosphate group and through the support of an organism's genetic heritage. Recall that our energy creates our structure and then our structure creates and controls our energy. This is the true nature of a healthy identity, for me the Naturvore identity.

Life's energy moves through space (locomotion)
and time (ancestry) as a current of power (ATP-based)
and a stream of biological information (DNA-based).

I find it remarkable how life's energy and forces actually flow through us. In fact, the biological energy created by the original organisms on planet Earth still flows through us because, once created, the current has never been completely extinguished. The power that our ancestors possessed is transferred to us and organizes us into life forms, a current of information that becomes our current of power. In other words, the Natural Order is preserved through ATP.

Although usually thought of as somewhat difficult grasp, biological energy can be measured in kilojoules just as the power that runs a light bulb can be measured. Because the energy transfers I have described are the result of phosphate bonds being formed and broken through ATP molecules, we can also measure the power of a positive life. It is easy to know how many calories we need to eat to sustain our bodies or the electrical energy causing a muscle or nerve to depolarize. Positive energy (benefiting the organism's life) can be measured, harnessed, and capitalized on, and therefore is not so mysterious after all.

ATP can move between cells in specialized pores in the cell membrane called gap junctions, and so it does not require typical, and more cumbersome, intercellular transport methods as most other biologically active molecules do. Instead, it moves similarly to water and small minerals. This biological energy permeates our tissues yet moves through us with properties that can be described as currents or fields of energy. We can even imagine a protective force field that envelops our tissues because the function and integrity of our skin and our immune system are dependent upon the power generated within and conducted through our skin barriers. When we lose the capacity to regenerate ATP, as in death, our protection is gone, but our skin tissue may appear intact to the naked eye for a while. We are somewhat deceived by this appearance of being intact because the most important barrier created through the metabolism of ATP is gone. The force field energy has been lost, dissipated back into the universe. As we lose our protective energy, we are unable to combat threats from toxins, the weather, and other organisms. We are at the mercy of the environment. Upon death, we become another growth medium for the organisms that devour us.

While alive, the energy derived from ATP can be summoned and directed to suit our needs. The movements of blood, nutrients, oxygen, electrical currents (depolarizations), and water are determined by the fluxes of ATP-derived energy created by our bodies. For us, efficient distribution of our energy means that these fields are able to penetrate the various cellular milieus in an unimpeded manner. As with any consideration of energy production and transfer, the integrity of our tissues (the conductor) will determine the efficiency and ultimately the power of our energy because our tissues are the very fabric that both generate and conduct the energy.

However, our power is made infinitely more valuable when it is aimed precisely, and this is the benefit of a well-formed identity (pun applies). We in effect create electricity, but most people will think of this like the mechanics of electrons along a wire, as stored electricity to be tapped for a purpose. However, as a laser beam is the transmission of energy in a focused fashion, as a lightning bolt is the transmission of electrons along a path between the sky and the ground, our precisely focused energy is much more powerful than as a source to be tapped at random. In short, biological energy associated with ATP (our energy particles) can be focused into more effective uses to achieve a maximum use of the energy available.

Individual organisms are easily seen as isolated bundles of energy, but they also create larger energy fields when they are acting in coordinated unison. For instance, dolphins are organized into pods and live in tight social circles in such a way that the group improves each individual dolphin's life. Each one benefits from the others' energy production. The pod hunts, defends, and plays as one unit. Even as dolphins sleep and the pod is semiconscious, the group remains coordinated and in constant motion. The pods I have seen in the Kealakekua Bay off the big island of Hawaii travel as an undulating wave, each dolphin swimming synchronously while the entire pod rhythmically ascends to the surface to breathe air and then descends to a depth of seventy feet for protection. In fact, the pod's activities exhibit a specified frequency of several minutes and an amplitude of roughly seventy feet, qualities representing a wave of energy.

The dolphins' pod is part of their identity, and we can observe the larger sense of it by watching them move through the ocean. Sea water conducts energy much more efficiently than air, and so, being creatures of the ocean, a pod of dolphins can be viewed as a coherent field of energy made up of individual particles of energy (each dolphin) undulating through space and time as do other forms of energy, such as light and sound. Prey and predators of the pod understand the capacity for power inherent in this wave of energy because the dolphins can strike like a bolt of energy. Like this pod of dolphins, individual humans are more efficient at survival when they too can focus their energy toward positive living. The energy made in cells creates a dolphin and then a pod. The microscopic is translated into the macroscopic.

Obversely, to lose the focus of one's energy is the first step toward entropy, toward death (the absence of coherent life energy). Disease results because the distribution and production of energy required to live by is disordered and therefore not able to serve the purpose for which it is intended. As discussed in "Cellular Corruption," the protection, repair, and healing mechanisms intrinsic to our bodies demand efficient energy production and efficient energy transfer. Our energy is meant to work at defending against microbial intruders and preventing cancers from forming and metastasizing in our bodies. Diseased body parts are by definition corrupted mediums for efficient energy transfers, and they can be likened to frayed power cords or downed electrical substations that cause regional power failures within our organs. When our bodies fail to produce, conduct, and project the energies required for an optimum life, we become limp and weak.

Diabetes, atherosclerosis, cancer, and the other chronic diseases impede efficient energy conduction throughout our body. After years of experiencing high blood sugars, diabetic patients lose the sensitivity of a refined nervous system and cannot sense an infection taking hold of their body parts. Diabetics not only become numb, but also tend to lose their capacity to direct their mortal energy efficiently enough to provide robust defenses against infections and cancers. High blood sugars directly injure the tissue matrices that produce and conduct the energy required for immunological activities.

Coherence of our energy fields creates biological strength, resiliency, and adaptability in our physical beings; and the best physical explanation of our power through coherent energy fields may reside in the theories of quantum mechanics, the physics discipline that attempts to explain the relationship between information, energy, and matter. Within our cells, the quantum forces of the universe apply, of course, for all biological activity, and the transfer of energy in general, is a form of information transfer, the quantum of consequences we experience. Likewise, philosophers of science, such as Henry P. Stapp of the Lawrence Berkeley Laboratory, believe quantum mechanics describes how consciousness can be created and sustained from the atomic structures of our brains. Again, energy is information and information is energy within our biological world, and so who we ultimately are is a function of energy transmission too. We are possessed by energy and the better "suited" we are to bring coherence to it the better off we shall be.

Quantum energy packets are creating our stream of consciousness and can be realized all the way to the banalities associated with a bicep curl during a work-out or the energy required to stuff Twinkies in our faces. Thus the body and the mind are, at the level of energy transmission, each one in effect a piece of us; and so the way we treat one in terms of inputs affects the other. When stuffing Twinkies in our mouths is part of our identity, we have lost the purpose of nutrition, which is to live with the most vital tissues and the most coherent energy pattern. Perhaps it is easier to understand that the same applies to addictions, wherein our activities and biological energy lack coherence. Pervasive compulsions, such as addictions, represent the ultimate hijacking of our mind's energy. All addictions consume purposeful energy, diverting resources to feed the addiction. Dependencies begin with and are reinforced by the change in brain architecture as a result of repeated exposure to the offending agent, whether it be food, tobacco, medicine, or other intensely gratifying sources. Once the brain's architecture is changed (by metabolic alterations), much of the addict's physical energies are diverted from productive activities to instead be invested in self-destructive activities. Our tissues dictate whether or not our energy transfers because they are the governors of our energy patterns, beginning at the atomic level. Incredibly, the transmission of energy within us depends upon a free radical or electron's angular spin that decides our fates in life. Our tissues govern the effects or tame these atomic particles.

Free radicals are omnipresent and simply part of the structure of the universe. These (generally) negatively charged atomic particles can be fashioned into a positive energy, depending upon the context of the situation. Free radicals are harmful to us when they are out of place, disordered, or exempt from the appropriate context of our life situations. The angular velocities of free radicals and electrons must be kept in check for the sake of our vitality, and we have the capacity to control these electron agents and use them to our advantage. There is often a certain propensity for randomness of free-radicals, for example, when the inflammation-induced unpaired electrons are made inappropriately, but truly vital tissue can bring them in line. ATP is regenerated by the oxidative potential of free radicals. Therefore the power of life resides in the electron transport chain of our mitochondria just as it does in the photosynthetic chain embedded in a plant's chlorophyll, the organic matrices that provide positive power by directing charged atomic particles (free radicals) into cellular respiration.

There should be little doubt about our capacity for and natural adeptness at harnessing the forces of energy and translating them into the meaningful events of our lives. As I bask in the sun, photons (light energy) will strike my face, and, combined with the thermal radiation from the sun, make me feel warmth. More interestingly, these particles of energy translate into biological energy and serve to elevate my mood, and do so very potently, even though the source is 9 million miles away. Throughout my body, the same photons of sunlight providing my happiness are harnessed to add a chemical group to vitamin D molecules, creating a more active form of the hormone that strengthens my bones. Again, biology is all about the capture and transfer of energy.

It is via the choices we make on a daily basis with impunity (especially those pertaining to our foods, medicines and physical activities) that energy management takes shape and, consequently, our bodies take shape. Choices of foods and activities determine our physical structure, our mental contents, and ultimately the biological context of our lives. The integrity of our cells is the context from which we turn something potentially negative, such as free radicals, into positive energy. This is indeed sanguine, a hope eternal that our fates are what we create using the positive power assigned to a natural life.

Corrupted cells are weak and cannot fashion the forces of the physical world to sustain a confident physical presence of health, and Americans are distinctly out of control of their energy management, as noted throughout this text. People's tissues are being molded by the Industrial Order on a vast scale, their bodies marching to the hum of industrial might and pushed into a state disconnected from nature. Moreover, there seems to be a widespread ratcheting effect that prevents our return to a more natural state once we have lost our way from the Natural Order.

I have heard it said that the brave die just once, but cowards die a thousand times. That being said, we must stand up to the Industrial Order and make the right choices that will preserve our true nature, perhaps making socially uncomfortable choices. Allowing the Industrial Order to permeate us is similar to a thousand acts of dying in that there can be a macroscopic appearance of normality before our tissues become objectively ruined, amputated or dead. By understanding the nature of our life force and our relationship to the Natural Order, we are less likely to be deceived by the semblance of normal tissue (that being sustained by the Industrial Order) preceding disability and death. Unfortunately, our young people are entering the dying process prematurely, but this need not be so.

Sickness tends to create more sickness, not only within us but around us as well. Through industrial medicine, American society places a special context on our aches and pains, raising their significance to grab market share. People often speak of the concept of "gateway drugs," but what about "gateway sickness"? In other words, our release from this ratcheting effect cannot be accomplished by applying more technology, and in fact our interactions with medical centers not only dissuade our return to a more natural disposition, but the ratcheting effect continues as we enter into a medical regimen and are subjected to so-called standards of care that have nothing in common with nature. Standards of care in modern medical systems debase our individuality and our creative spirit, from which we can all eke out a healthy existence, if our innate and authentic biological energy is not corrupted.

As noted many times herein, the integrity of one's life is dependent upon the integrity of one's tissues, but a more accurate statement would be that the integrity of one's life, including one's notion of reality, is ultimately tied to the integrity of one's energy fields, which are a result of ATP. Life may require water as well as energy, but water does not make things happen as energy does. Indeed, the purposeful transfer of energy is the most fundamental property of all life. In "The Ethic of Self-Preservation," I explained that one's feelings and physical symptoms can have either a positive context or a negative context, depending upon the state of one's physical condition at the time of the symptom. This is curiously similar to the energy context for all life, wherein one has a relationship to the atoms of the universe and the associated free radicals from those atoms. The organization of one's atomically-derived energy is how one resists the predisposition to entropy and disease. As we should expect, energy management at the macroscopic level of emotions and choices is translated into subatomic energy exchanges.

In other words, our connection to the physical laws of the universe is maintained through the integrity of our humble tissues. We are fields of energy that are in effect organized by our tissues simultaneously both macroscopically and subatomically. By reading these words, the readers' energy fields (especially in the brains) have been altered through a series of biochemical reactions that create meaning as chemical energy is translated efficiently to electrical energy. Subatomically, the memories of what has been read represent the purposeful use of electrons inside and outside the readers' brain cells. Emotions and other durable effects can be woven into the readers' brain-architecture, the fabric of the person, despite the effervescent qualities of these energy events. One could also say that the choice to read itself is yet another aspect of energy management. The **desire** to seek a healthier life is a type of energy transfer that led the readers to my words and the subsequent reactions in their brains' energy. Health is inspirational, the flow of positive energy from one life form to another, whether it is derived from who we are with, what we do, or what we eat.

Within only a few decades, however, the contextual concerns of our personal energy management have undergone dramatic changes, making this choice to live a healthier life rarer than it ought to be. Barry Popkin, PhD, a professor of nutrition at the University of North Carolina (UNC) at Chapel Hill, along with other researchers at UNC have found that there is a global decline in the routine physical activity of humans. The decline is most pronounced in the industrialized countries but is now taking place in the developing countries as well, a trend that portends an ominous fate for the species.

> Using a physiological measure called metabolic equivalent of task (MET), which describes the amount of energy spent in accomplishing a task, the study determined that by 2020, the average American adult will expend about 190 MET hours per week.
>
> In comparison, a person who slept 24 hours in a day would expend 151 MET hours per week, and an active adult who did vigorous activity for 30 minutes to an hour every day, but otherwise had a desk job, would expend between 240 and 265 MET-hours per week.

(*Obesity Reviews* vol.13, issue 8:659-680, Aug. 2012)

In effect, we are a people who appear to be actively engaged in life, but in truth our energy signature approaches that of a person who is sleeping full-time. Although not mentioned in the Popkin study, it is noteworthy that the adoption of a less energetic lifestyle coincides with the adoption of more manufactured foods. This is the perfect storm for incoherent energy patterns within us.

The Naturvore lifestyle provides benefits that are both microscopic and macroscopic. On a microscopic level, I have improved my biochemistry, which, at the macroscopic level, has improved my overall physical performance dramatically. But there is also a great deal of benefit in the alignment of the natural energies and the reduction of entropy. Improvements at the microscopic level translate to a physique that is more consistent with the wild state of the human, one built for strength, speed, and endurance on a macroscopic level.

Being a baby boomer and a physician, my acceptance of the Natural Order as the governor of my most vital affairs is at best improbable. Some would argue that it is a bit "whimsical" to accept such a premise, that the Naturvore identity holds the key to health, but I sustain my conviction because I am frequently reminded through my work as a physician that detachment from the Natural Order puts our lives in jeopardy. The initial resistance I had to fully implementing the Naturvore principles dropped precipitously as I began to sense increased personal power through the changes in my lifestyle. My resistance to the Natural Order declined further as more personal power was realized and the positive reinforcement continued within me. Deep reflection upon my improved capacities and changing circumstances engaged a strong compulsion to write this book.

I find it extremely fascinating that our individual integrity as people as determined at the subatomic level can be seen in our integrity within larger socio-economic affairs, and vice versa. Many people seem to have a healthy respect for how tranquilizers, narcotics, alcohol and tobacco influence our brain addiction pathways, but few people acknowledge the influence of breads, pastries, cakes, candies, chips, Big Macs and other manufactured treats on brain energy patterns. The fact that many Americans repeatedly consume foods that directly contribute to fat-filled arteries, weak hearts, weak brains, and fatty livers, foods that alter their physical identities in such a way (obesity) that it alters the content of our life experiences, proves my point. Anything that can alter our brain wave energy patterns by changing our neurotransmitter milieu can alter our visceral energy, our vitality, and our spirit of life.

Conversely, proper energy management that results in a physically trained, properly nourished individual (the most authentic biological organism) is the ultimate defense against the subliminal onslaught against our health in modern times. The Naturvore identity I have achieved is in fact a physical presence created out of the energy generated by the mitochondria in my cells but channeled into a coherent pattern by physical and mental congruence. The energy emanating from the mitochondrial organelles throughout all of my tissues diffuses through my body, but the effect is like an energy field that aligns the proteins, the genes, and my thoughts both in terms of intent and my attitude. My intent is to make my life real and really mine, and my attitude in this pursuit is one of vitality to match my physical vitality. In other words, the structure of our identity boils down to the energy we create and conduct, and the more vital that energy, the more vital our identity.

A unified body and mind is intended by nature, is of the Natural Order, while the Industrial Order diverts our energy to serve its needs by dividing the mind from the body and setting the former against the interests of the latter. The Industrial Order makes it seem reasonable to consume unhealthy products rather than the health-optimal products of the Natural Order, which in turn further misaligns our energy patterns creating less resilient tissues. The good news, however, is that this is reversible, that the ratcheting effect works in both directions: the Naturvore identity reinforces itself because the positive energy of the lifestyle creates resilient tissues and a resilient mind, which then can better withstand both the Industrial Order and disease.

As proof of our energy management as the best method to secure a healthy and happy life, let us consider the opposite by reviewing one of the most common forms of disease and causes of death in the United States, atherosclerosis. In the Industrial Order, plaque in our arteries is often referred to as a sort of blemish, a nuisance of the modern age, one that can be polished away by the routine ingestion of pharmaceutical drugs or perhaps pushed out of the way by a cardiologist's metallic stent, which is frequently left inside the patient as an artifact of our modern lifestyles. These are industrial machinations, the end-organ results of atherosclerotic disease, but these clever innovations and the science that provides the means for their deployment fail us on many levels. Atherosclerosis is not simply a plaque or a clogged artery, but a complex process that symbolizes poor energy management throughout multiple planes of body tissues and multiple dimensions of our lives. One dimension relates to the disease process as a matter of choices, and, although atherosclerosis may result in clogged arteries in its more final stages, more often than not it is a reflection of one's values in life and one's energy management style. Atherosclerosis is the *sine qua non* of Industrial Order diseases because it is a common result of poor energy management.

The manufactured food diet promotes fat-deposition and inflammation, clear signs of poor energy management through poor nutrition, but this situation is made more deadly when combined with a sedentary life. Americans are also routinely attached to and dependent upon widgets and gadgets that make life easier and thereby reduce energy expenditures. Reduced energy expenditure means that our tissues make fewer energy demands on our closed circulatory system. Our bodies are born to be efficient, and therefore energy demands (activity and metabolism) are matched closely with energy delivery (circulation): with less physical activity there will be less blood flow to the tissues. Simply put, the sedentary lifestyle requires less blood flow because it has fewer energy demands, and this happens years before atherosclerotic plaque has begun to develop. So, an apathetic mental state causes physical disuse which reduces the energy demands placed on the heart and arterial systems long before the atherosclerotic plaque is forming. The hardened and narrowed arteries the cardiologist's angiogram ultimately reveals is a result of nature taking our original arteries away from us because they are not being used as intended. In nature, all physical attributes that we do not use often or do not use optimally tend to be lost, proving the adage "use it or lose it."

Atherosclerosis is an energy transport problem rather than a plumbing problem, which is how Industrial Order medicine treats it. True, beautifully natural arteries are being replaced by less functional spongy, foamy, plaque-filled (anyone care for a cream puff?) semblances of real arteries, but corrupted arteries cannot carry the energy of life efficiently, and that is the cause of suffering for those with atherosclerosis. Medical, considered "optimal" by healthcare providers, management of atherosclerosis fails to acknowledge the importance of energy management in the creation of atherosclerosis. We create atherosclerosis, and this is why it is happening to younger and younger people. The energy-flow problem needs to be fixed, and just doing plumbing interventions is not going to get the energy flowing properly but only sustains sub-optimal energy conditions for a time.

Viewed from another angle, maybe atherosclerosis is simply an immunological attack, the inflammatory cascade being one component, upon the architectural distortions resulting from the cellular corruption (industrially-created, artificial entities themselves) that is sustained by artificial foods and the sedentary life. Naturally refined immunological systems are provided coherence by the same energy management responsible for all of our life functions, and therefore discordant energy utilization translates into immunity without regulation. The arteries we are intended to have can actually be dissolved by inflammatory cascades, a process of autophagy that is aligned to the natural aging process as well as to disease. The devastating effects of atherosclerosis can be felt throughout our bodies, within our heads (strokes), our gut (ischemic colitis), our genitals (erectile dysfunction), and all the way to our feet (peripheral vascular disease).

Atherosclerosis is an industrial process that often leads to sudden death, but the disease is a practiced affair, beginning with a loss of coherent, naturally purposeful energy flow. Industrial-made routines tend to train our energy and eventually extinguish our life energy. Indeed, industrially aligned energy management does not just make us susceptible to atherosclerosis but to other maladies, even cancers, as well.

Vigorously avoiding the industrial poisons that are apt to destabilize our DNA and thereby make it prone to cancerous transformations is the first step in preventing cancer. As with atherosclerosis, energy management that creates resiliency to cancer begins in the brain as a thought or intention but can be transmitted throughout the body subconsciously (as in a seamless tapestry) through coherent energy patterns. By committing our lives to cellular health, we maintain the tissue matrix nature intended us to have, one that efficiently transfers energy to resist cancerous transformations, detects cancers if they occur, and mounts and directs a response to subdue the cancerous insult promptly. Whereas atherosclerosis could be viewed as a new malady born of modern living, cancer is an old problem in evolutionary terms, and therefore our bodies can be expected to have exquisite natural abilities to terminate cancer, but we must cultivate and hone those mechanisms (which the Naturvore lifestyle accommodates nicely).

However, there is an Industrial Order twist that makes cancers a modern disease too inasmuch as they arise out of tissues that have lost their real biological purpose, wherein the architecture of the biological fabric (tissues) has been liquefied, the cellular architecture ravaged by chronic inflammation (this includes the changes from using tobacco products as well). Avoidance of inflammation can be accomplished by simple measures such as eating a calorie-appropriate diet and foods containing abundant antioxidants to absorb the chaotic free radical energy particles that are associated with cancer and aging. It has been shown that the infusion of highly malignant cells directly into the bloodstream of a lab mouse does not necessarily cause cancer. The biological fabric of the mouse must be sufficiently degraded for the cancer to take hold. Even the most metastatic cancers must find a favorable spot to call home, which means the maintenance of optimal tissues is perhaps more important than the avoidance of carcinogenic exposures.

Our immunity must remain strong to defend us against foreign invaders and cancer, but our immunity must also remain ordered and purposefully restrained to prevent autoimmune diseases, such as rheumatoid arthritis and lupus. A biologically purposeful diet and exercise plan not only build the most successful defenses, but also tailor our energy fields to exact the most meaningful use, our optimal life. Governance of our energy fields means precision with ATP transfers, which translates to higher order intracellular and intercellular signaling mechanisms, which means we are less likely to experience auto-immunity and intrinsic self-harm.

THE POWER OF POSITIVE THOUGHTS

Obviously, the integrity of our tissues is of paramount importance for our current and future health, and only through natural means can we sustain the best tissue integrity possible. However, there are aspects of the efficient transfer of biological energy that results in a synergistic relationship of body to mind and vice versa that have yet to be explained by science, aspects that can have a pronounced effect on our health. The use of a placebo has been shown to substantially lower the mortality rates for all diseases, even if a disease is well established, which suggests a seriously important relationship between mind and body (I recommend Harvard professor Anne Harrington's *The Cure Within* about the history and power of spiritual cures, positive thinking, and the placebo effect). The placebo can effectively treat or cure 20–40% of cases of almost all known medical conditions, and yet science cannot explain why this is so. Incredibly, it seems that something as simple as faith in a lifestyle, pill, or other device can assist us in focusing our energies on the cures we have within us.

During the early 1990s, an orthopedic surgeon named Bruce Moseley tested the placebo effect in his surgical suites. Using a standard double-blind study, he enlisted two groups of patients for arthroscopic knee surgery. He enrolled these patients through his office and would only open the envelope to determine to which treatment group they belonged while in the operating room. Both groups of patients were taken to the operating room, then prepped and placed under anesthesia in the usual fashion, but for those in the control group, he performed his usual arthroscopic knee surgery, while for those in the other group of patients "Moseley actually did not cut, scrape, or do anything therapeutic to their knees at all — he just opened them up and then closed them again" (Harrington, 2008). Both groups had successful outcomes. Those patients who had received the placebo surgery (not undergoing traditional surgery) recovered with improved knee function, experienced relief from the symptoms that had urged them to consider the surgery, and were just as pleased as those who underwent the surgery (rates of success were similar). In fact, many reported symptom relief years after the sham surgery.

The placebo effect, which obviously has to do with the transfer of energy at some level, most likely through the mind focusing energy on self-healing, has been reproduced consistently since the advent of the scientific method under a myriad of conditions. In fact, the placebo is so demonstrably powerful that there has been a renewed interest in its effects, but less so within the traditional medical arena. Perhaps it is the fact that physicians cannot own the placebo effect that makes it a less tempting target of inquiry. There is also the possibility that we are not sophisticated enough to understand the placebo phenomenon, even though it is something very elemental. Indeed, despite its universally acknowledged power and potential for benefit, physicians are admonished if found using it on their patients, and all of the major medical societies have made very clear in numerous position statements that the placebo has no role in treating illness, even though the placebo effect is the benchmark from which to measure all newly purposed medical interventions. However, the truth is that we do not need a doctor to tell us something positive in order to focus our energy on our healing. It would seem that the placebo effect is elemental to the human species, and an individual can benefit from it without entering the medical-industrial establishment merely by proceeding with a purpose of mind and body: to be in excellent health.

THE CHALLENGE

The Naturvore identity provides me resolve for the most powerful tension, one that operates both at a conscious and a subconscious level in my life: the Natural Order versus the Industrial Order. However, the tension seems almost ludicrous when I consider that the Industrial Order lifestyle involves chronic exposure to toxic materials, which means the only way to resolve this tension satisfactorily and most completely is to live within the Natural Order, where the positive energy provides immunity from the chronic toxic messaging of the Industrial Order. I don't mean to understate the difficulty of doing this, however, for successful alignment to the Natural Order is an accomplishment of immense magnitude during these modern times.

While visiting a very dear friend convalescing from a near-death experience recently, I enjoyed a five-mile run that provided me a reflective moment. Midway through the run, I arrived at the John Brown University in Arkansas and thought about the thousands of young minds therein. Although it was winter, I thought it odd that there were only a few students outside, and certainly none was exercising.

I then found a cemetery next to the John Brown University. I had never run through a cemetery, and so I checked it out. While running amongst the headstones, I felt seriously engaged in a physical life, the graves being reminders of how we are organic, physical creatures and that life is absolutely magical. The alternative seems to possess no magic at all. It is interesting to note that dead people and living people share the same material constituents for a certain amount of time, the only difference being that the living are still generating ATP molecules and the force of life while the dead are not.

Although this run was in and of itself physically important to me, the real benefit gleaned was some time to get my thoughts together in order to help my friend Larry best. He had survived some serious wounds, a blood clot to his lungs, and a massive heart attack, which together had caused him to go into cardiac shock and respiratory failure, and he needed a good dose of positive energy. He had been placed on life support several days prior to my visit, but Larry had taken good care of his body before the event and so had good reserves, and he was already at home making a marvelous recovery when I arrived. Larry and I spent time together reflecting on how life is indeed quite fragile. Larry told me he had been given a new opportunity and expressed that he had a new respect for his biological life.

After my workouts in Siloam Springs, I felt a sense of synergism. I not only was maintaining my own health but felt that I was also improving my friend's. We did not do much sightseeing, as he was still recovering from his ordeal, but we did spend time in the grocery store together, and I was able to show him how I shopped for food. Upon coming home from our shopping excursion, we had to make some arrangements in his cupboards and refrigerator. He reported to me that foods that he ate frequently before his medical emergencies did not carry the same appeal afterwards.

As a Naturvore, synergism, defined as two or more biological processes in alignment for a common beneficial outcome, is harnessed routinely. The results of these two processes aligning are not additive but exponential. Feeling the benefits of an accomplished exercise feat is good, but then feeding the body the foods that are optimal for recovery and health in general yields synergistic benefits. Healthy people eating natural food is better for the entire planet because there are fewer factories making foods and fewer factories making medicines for sick people. Remember: food factories breed not only medicine factories but also hospitals, medical plazas, and now medicalized cities such as Rochester, Minnesota (Mayo Clinic), and Cleveland, Ohio (Cleveland Clinic).

Being physically fit tends to change our life experiences to encompass more vitality and better experiences in general. In fact, I tend to consider the need to make it quickly to a medical center a few blocks from my office an opportunity for physical activity rather than an imposition. Just the other night, for example, my 14-year-old son and I went to an Elton John concert. Parking was difficult near the arena, and so we parked several blocks away and across a highway bridge from the entrance. During the concert it occurred to me that our truck was squeezed into a corner space far from the parking lot exit, and my son and I would have to wait for the entire parking lot to clear before we could exit. This was a dilemma because Elton John played until almost 11:00 p.m. and I needed to be in the hospital by 7:00 a.m. We stayed to the end so as not miss any part of Elton's encore, and then we bolted from the stadium at a good 7-minute-mile pace. We were careful among the pedestrians, but the streets were blocked off because of the concert and so traffic was not an issue, and my son and I experienced the excitement of running through Asheville on a chilly, damp, and windy night. People were moving along as if in a herd, and we were just breezing by them with the wind in our hair. While running over the highway overpass, we saw lanes full of cars whisking below us. We arrived in time to pull out of the parking lot before anyone else had reached it, and I noted how the spontaneous run added to the concert experience.

My achievement of a Naturvore identity transcends time in a number of ways. For one, I have turned back the clock to large degree, forestalling the diminishment of physical function that comes with aging — and definitely forestalling the effects of aging predicted by the Industrial Order for someone my age. I am also more aligned to my ancestors genetically, which is made possible through enhanced DNA-reading within me. My physique also appears less "civilized" because it is not cloaked in the typical layers of fat of someone from an advanced society, and metabolically my diet is consistent with more ancient times. I am also linked more sincerely to the present, however, demonstrating a respect for my immediate biological moments. I am also constructing a better future through the biological memories I accumulate, biological moments more likely to produce a healthy longevity. Time on this planet has taken on a new meaning and a new reality for me. I intend to be more prepared for the challenges of the future, but this means I must be consistent in my actions.

Several times a day, I bring nature into my body for nourishment, and I never feel sluggish after a meal or corrupted in any sense. I engage in running, weight-lifting, and other physical activities not so much for the caloric expenditure but with the intention to awaken my physical presence. My workout partner, Mark (the Batman), has shown me the value of warm-ups as a focusing activity. Preparation for a workout provides better results because the mind tends to be more present for the activity and the muscles and mind are beginning to be unified before the work begins. I prepare myself for a great performance even if I am not in a competition. I want to feel my workouts for what they are worth, and therefore I focus my mental energy during these important biological moments. In a similar fashion, I practice mindful eating largely to enjoy the infusion of my nutrients and to marvel at the beauty and satiety nature can provide. I make adequate time for my lunches and snacks, and I tell my patients and friends that this is a sincere form of self-respect. I tell them that we should all take the time and make the effort to improve our performance throughout the day, that our experiences are largely what we make them, that they don't just randomly happen. In other words, I wish to become more adept at focusing my psychic energy on what is real and what is healthy in order to live to the best of my ability. It was this realization, that the best defense against unwanted intruders into my mind is a well-formed identity, which will improve the contents of my mind and therefore my life, that led to this notion of the Naturvore lifestyle in the first place.

I also practice meditation, at which I am no expert (why is it so hard to think of nothing?) but I know that it can center us into our bodies and offers existential peace. When the mind is quiet, the body and mind are unified because the mind is not trying to run the show; but I suspect that there is much being done within my body while meditating because it is generally more capable of house-keeping chores and recuperation if not distracted by the mind. Certainly, below my immediate awareness, millions of neurons are quite busy sensing injury, guiding immunological responses, and coordinating the refurbishment of body parts.

Meditation allows us to focus our psychic energy, the metabolism of ATP molecules associated with brain activity, to the simple aspect of just being. In fact, in meditation, we are not doing, we are just being — as in being human. Perhaps there are benefits such as those associated with sleep, whereby my memories can be processed into a coherent picture for better use. Meditation is also, metaphorically, like cleaning up a computer by purging the mental remnants of the Industrial Order, the mind cookies, memes, and corrosive junk files. I also get the impression that, while meditating, I am consolidating random access memory banks for improved mental efficiency.

Our mental lives do not occur in isolation from our physical lives because the body is the sensory structure for the brain and supports the brain's function. The brain in turn governs the responses of the body as it engages with the world. The brain and the body are, in short, one entity without borders because, without a body, the brain has no ability to construct its version of reality. Therefore, as we improve the integrity of all our tissues, they become more adept transducers of the energy necessary to experience our lives.

Within the Naturvore identity, I remain aware of the physical composition of my life. Purposeful strenuous activity induces growth, adaptation, flexibility, and versatility within my tissues. Occasionally, I catch myself asking if I really need a workout today. I have a million and one excuses, of course, but I ultimately ask myself, "What else would I do with my time that is better than a workout?" That is the crux of the matter. I have found strength and resiliency in a unified mind and body, both geared to improve each other and the biology I was born into. Coherence between my intentions to have a healthy life and the actions of a healthy life creates both sanity and the most favorable physical experience. The energy fields produced and projected by my brain are synchronized to the energy fields produced by my body. Coherent energy produces a coherent person. Therefore I cannot think of a better expenditure of time than to honor the intentions of a healthy life with the actions of a healthy life.

Then, as I stretch, perform push-ups, sit-ups, and other calisthenics, I am always grateful for the gift of another day, courtesy of my body. I refer to these activities as my "breathing exercises" because nature dictates that I must earn my right to breathe today. I perform this simple routine in the morning, a time when societal pressures mount rather quickly and begin pulling at me, trying to eviscerate my singular purpose, the essence of my life. Nevertheless, this is what I have chosen to do, and the demands of the day, which I will meet with all the more vigor for the decision, will wait until my morning "breathing exercises" are complete. The decision is, in fact, simplicity itself. I have found that preservation of the self means just that: simple measures provide less distraction in life and yet often create the most exquisite life experiences.

TRANSCENDENCE

"The power which resides in him is new in nature,
and none but he knows that which he can do,
nor does he know until he has tried."

— Ralph Waldo Emerson

As I work out at various gyms, I am often asked what am I training for, which I assume is because the people asking are curious about my physical form. I have achieved the physique of a competitive athlete, even when compared to people half my age. The question gives me a generous dose of satisfaction because the people asking tend to be in their twenties and training for sports competitions, but my answer is straightforward: I am training for my life.

Many of the basic assumptions regarding what to expect out of life tend to be provided by, and ingrained into us, by our culture. Consequently, before I realized the Naturvore identity, I thought we were to be less physically fit and less physically capable in middle age than we were in our youth. I have found this to be a false assumption. It has become obvious to me that muscular atrophy and obesity are not inherently "middle age" physical qualities but assumptions we all make based upon the way we are encouraged to live, the foods we eat, our inactivity, and our submission to technological medicine. Indeed, if we believe these qualities to be inevitably the case for us in middle age, then they will be true. In other words, we can all become sedentary middle-aged people with chronic medical problems if we choose to, but indeed we have a choice. My current physique and physical abilities are proof that we can transcend our culture's definition of our chronological age and escape the grasp of the Industrial Order.

I serve as proof that it is not necessary to be perpetually lulled into poor physical expectations of ourselves or to eat the standard American fare that corrupts us physically, that it is not necessary to live at the mercy of the medical industry, gasping for life and grasping for hope within a life context that consists of artificial life support.

We can arrange better expectations and better circumstances beginning with the fact that we need not be who we are now; we can be better.

I want to know what I am capable of and what I am supposed to be like physically, and therefore a compulsion for self-improvement has become a distinct focus of my life, even at the age of 48. Because I have changed so dramatically for the better and have even surprised myself (the doctor who knows so much), I feel it is essential to continue on a path of self-discovery. I fill my days exploring my physical potential and creating new abilities because there is one thing for sure: I have yet to know all there is to know about myself as a living organism and what I am capable of.

Perhaps one would think that, after being confined to this body and this mind for forty-plus years, I would have a fairly good idea of what I am capable of, but this is not the case at all. I am still learning a great deal about myself: my desires, my dislikes, and most important, my intrinsic capacity for life. In order to routinely explore my performance and mental and physical potential, I must remain busy doing it. Biological competency is created, while biological authenticity is chosen. It is through our choices that we explore ourselves and our world. Actions always speak louder than thoughts, and I can say that I underwent a profound transition in my life through my simple, very personal actions, less so through pure contemplation, and not at all through engagement with a society that is focused on what it can do to me. But as the reader will find out more specifically, this simple transition yielded a better translation of me, my biology. The changes in my life, including the ones in my mind, are only possible with a different translation of my DNA and the other physical elements of my body. My life continues to unfold, but not as if these are new chapters in the same book but as an entirely new book, complete with a new binding and a new cover. As it turns out, my life is unfolding differently just as a biochemistry researcher sees that one protein folds and unfolds three-dimensionally differently from another protein. Our lives are physical, therefore the products of our bodies are much more than we usually consider them to be. We can improve our minds' functions by improving our entire physical form, today and at any age. Brain disorders, such as strokes and dementias (brought on by obesity and diabetes) are caused by a failure of the entire body; they are caused by the same cellular corruption that erodes our entire physical nature. A better body translates very efficiently to a better mind and a better life.

If we care for ourselves and protect our minds, bodies, and vitality, then we will also be in a better position to protect other things of value around us. In fact, the things I value the most have changed with this new body. For instance, instead of so much value being placed on the accouterments and materialism of an industrial identity, I am more interested in aspects of my biological identity such as creativity, curiosity, resiliency, speed, endurance, strength, flexibility, hope (that springs from an improved condition), spirituality, and even gratitude.

INSANITY

At one time in my life, I had very low expectations for my physical function because I was measuring my success by the standards set forth in the Industrial Order. My body's physical function did not seem important to my success in life, and therefore it was easy to reduce the expectations I had for myself physically. Reduced physical expectations were inevitably accompanied by reduced physical demands placed on my body, and I led a sedentary life like that of most Americans. In other words, the lack of physical expectations we place on our bodies is a cultural "value," but it became my personal reality. As night follows day, a disordered energy balance must ensue as we take in too many calories and expend too few. This is simple arithmetic, but the complexity of the problem lies in our interactions with our environment and culture, which becomes much more complex as we consider the conditions of the various elements of our relationship to the Industrial Order. Simply put, we tend to create the complexities in life that deter our health.

Through my journey I have realized an important consideration for our modern lives: we do not need to be physically fit to survive or to be successful people within the Industrial Order. Therefore our physical dissolution can occur without notable concern or indeed awareness, in spite of the insidious results unfolding before our very eyes. Since young adulthood, I have always been well versed in science, but that did not prevent lowered expectations for myself from creeping into my life. I became physically and mentally complacent as I plodded along, pursuing a successful life, enmeshed within the Industrial Order. It is interesting that we can convince ourselves that we are successful people, even as our organs are being destroyed.

Americans tend to become oblivious to the slowly evolving physical changes that represent a less optimal and unhealthy biological state of both mind and body, but this is almost expected because we have no natural predators, at least not those that threaten us in an immediate sense. Insidiously, however, we become prey to the Industrial Order as we place more value on comfort than on our health and become increasingly complacent. Before becoming a Naturvore, I was as guilty as anyone, not fully realizing that my biology is the mechanism for my life, in spite of my background, my vocation. I had simply become civilized, and with that I was not interested in self-preservation, except in the sense we associate it with our dog-eat-dog world of the Industrial Order.

In nature, the golden rule of exquisite biological efficiency that enables the survival of the fittest is "function creates form and form creates function." But in the Industrial Order, we are numb to the adverse changes within our bodies and ignorant of the missed opportunities to change the course of our health. My beliefs and the context of my life were provided by the Industrial Order, and therefore it is not a mystery why I would lose my allegiance as an organism to the Natural Order and to my very own biology. There is nothing difficult or particularly mysterious about this process, which amounts to a lack of self-respect, though we do not recognize the change in us as such.

As I changed my own expectations for my health, aligning them with the Natural Order, I began to see other people's expectations in a different light. I often hear patients say intently, "What are you going to do for me, doctor?" I sense they believe that health will come from their transaction with me as if they were purchasing a TV set, a computer, or some other routine consumer good. They expect me to "provide" them health from our interactions. In fact, I am called a licensed healthcare provider, but can I really provide them with health? The most intriguing aspects of these encounters is that, upon thorough inspection of their life habits, I see that they expect health to be delivered to them through me rather than creating it themselves. To their minds, medical centers, doctors, and other healthcare providers will matter more than anything they can do for themselves, including making a commitment to a healthy lifestyle.

Health is about our performance in matters of health, and to put it bluntly, it is insane to think that we need not perform as healthy organisms in a natural way in order to achieve health. To own our health means that we match our intentions to our actions, a condition that is extremely beneficial from a biological perspective in that we thus are aligning our biological energies to secure the best life possible. Recalling my previous discussion concerning energy matters, we should not underestimate the seamless nature of our tissues conducting our vital powers. It is perfectly sane to expect excellent health when we are engaged in doing all we can to secure it, but it is insane to not be engaged in health performance and yet expect to be healthy.

I find that as people move further away from the Natural Order, their expectations for health become increasingly unrealistic, a major cause of disillusionment. That is, many Americans do not seem to understand why they are not healthy, despite all of the innovative technology surrounding them. This disconnect from the Natural Order might be termed biological complacency, a tendency to the denial of the biological nature of us all in favor of a technological version. Moreover, however, people tend to be "broken," corrupted as regards both physical form and purpose, which allows them to believe that physical form and function are properties completely separate from their lives, perhaps worthy of a different, less serious consideration.

Disillusionment is not just a terrible feeling, in other words — it destroys health. There are many contributions to our disillusionment from the Industrial Order, but I will only focus on the ones that affect our health directly. As I have said before, we expect the medical-industrial complex to provide us health, as if it can do that better than we can ourselves. This emotional and physical reliance on extrinsic means to secure a healthy life is prone to fantastic disillusionment, and it is almost never addressed as such.

One way to see the magnitude of the problem is to consider the difference between health outcomes and the expenditures for healthcare compared to other industrialized countries. According to the United Nations Department of Economic and Social Affairs, the US ranked 40th in the world (2010) with respect to average longevity and, according to the Commonwealth Fund, ranked last out of the seven major industrialized countries measuring efficiency, access to care, patient-centered care, quality and ability to lead long, healthy lives. The US pays more than double the money per person compared to other industrialized countries to achieve those dismal statistics, spending $8,400 per person per year. Recent studies also indicate that more than one-third of the total money ($750 billion) spent annually on healthcare is blatantly wasted on administrative costs and therapies that do not improve the patients' health. By unemotionally evaluating the discrepancy of innovative, evidence-based medicine and our physical health statistics, we can clearly place a value, a measurable quantity, on our disillusionment.

Medical technologies and medical care in general cost a great deal of money, but do very little to improve population health, as pointed out by Canadian population health scholars R. G. Evans and G. L. Stoddart. This fact is also self-evident since the largest growth of man-made disease (the chronic diseases associated with American living) has coincidently occurred with the explosion of money spent on medical technology and medical care. Obviously, Americans believe in their technologies to a fault. The more we spend on diabetes care, the more people become diabetic and suffer the complications of diabetes, and yet it is a preventable condition. Prevention is far superior to treatment, and there is no technology required to accomplish this. Furthermore, the more mammograms are performed, the more breast cancer is diagnosed, but yet the death rates from breast cancer stay the same. We should all feel very disillusioned by our relationship to technology, especially in modern medicine.

Recently we saw emotions running very high and fear on full display as women were told publicly that the use of mammography should be decreased because it is thought to cause far more harm to patients than once believed. Medical experts had reasoned, as scientists do, that routine mammograms under the age of 49 years cause more suffering than they alleviate, but indeed the announcement itself caused suffering. But there is a false sense of safety with mammograms, and perhaps people should be more concerned about receiving too many mammograms. Very heated public discussions recently arose when the FDA recommended against the use of the chemotherapeutic drug Avastin (Genentech) for metastatic breast cancer since the drug caused serious harms without actually showing any benefit at all. Many women broadcast passionately that they had benefited dramatically from Avastin, but those women may have been some who had been over-diagnosed with breast cancer, meaning they may not have had it in the first place. Men have not been spared from a profound disillusionment with technology either. Profound consternation, disbelief, and a general unease came over many men with the recent recommendation by experts to forgo all prostate screening (PSA) tests because, after decades of widespread use, it was determined that these tests result in more harm than benefit. The uproar in all these instances seems to indicate that people generally appear very passionate about the medical technology available to them, that in fact they value their access to technology more than they value their personal fitness and self-care.

I have seen many patients look very pleased about new medications I have just prescribed or a glucometer machine for diabetes testing that uses less blood for sampling or the fact that there is a new open MRI scanner that will accept all body sizes, not just those less than 300 pounds. In fact, a recent billboard in my town advertises the arrival of an MRI scanner for all body sizes, yet one more indication of both the bad habits of Americans becoming the norm and of our reliance on technology for health issues. Furthermore, our dependence on technology often becomes another source of disillusionment because we can become victims to the need for more and more invasive tests to follow-up on things found in the original tests that have little bearing on our longevity or quality of life. The follow-up testing quagmire and treatments that often result can kill us. According to the New American Foundation website in 2008, estimates by Elliott Fisher, MD, a noted Dartmouth researcher, indicate that unnecessary medical care causes as many as 30,000 Medicare-recipient deaths annually, which approximates the total number of deaths caused by motor vehicle accidents in 2010 (US Census data). We are better off being resilient creatures, more confident in our personal powers rather than gullible to the advertisement distractions for newer technology, because it is more likely to let us down than build us up.

I came across a letter written to *Time* magazine, to Dr. Sanjay Gupta in particular, that clearly illustrates the unrealistic expectations people have with regard to medical technology. Dr. Gupta expressed concern (as I have as well) about the new societal push to love yourself as an obese person, which then promotes tolerance for poor health habits. An irate reader from New York opined, "Until the medical community can offer people effective solutions to obesity, I suggest that Gupta not knock the self-acceptance route" (*"Time* Inbox," Nov. 3, 2008). I would respond to this letter writer this way: self-acceptance should begin with the need to be more biologically authentic. We can hope for more medical care, or we can take better care of ourselves and be healthy. Obesity and the most prevalent chronic diseases in American culture need not be aspects of our very selves.

Unfortunately, our unhealthy expectations are instilled very early in life. Children are learning to accept disease, and even to expect disease, because they are more routinely visiting doctors and being placed on medicines. Children are also shown that living with disease is part of the stream of life via the omnipresent pharmaceutical advertisements on their televisions or just in the conversations they overhear adults having. It is not unusual, for example, to hear mothers exchanging information about new medications they are ingesting and new diagnoses that they recently received. Rather than accepting our diseases, we should be instilling the notion that it is preferable to avoid disease in the first place.

As noted, losing something as precious as our health and fitness by failing to take responsibility for our own health is insane — interestingly, the word "sane" actually comes from the Latin word for "healthy," *sanus*; around the sixteenth century, the word became more widely used to specifically indicate mental health and stability. Indeed, without sanity we are hardly healthy, and, as I have pointed out, most of our physical ailments could really be ascribed to a consumer-minded insanity. Sanity implies rational thoughts, a keen sense of self-awareness, and a capacity for honest self-appraisal. Indeed, self-preservation is the premier prerequisite for sanity, which seems a pretty simple assertion but does not apparently occur to most Americans.

As we descend further into our delusions, our insanity made more acute by cellular corruption, Americans become victims. The medical industry is in fact quite adept at making Americans the victims of forces in the universe rather than the masters and commanders of their fates. Blood sugars or cholesterol and other laboratory anomalies are usually blamed as causes for why people feel bad or have fatal events. I often see in the medical literature that aggressive interventions are acceptable when "diet and exercise have failed." The patients are never to blame for all that industrially produced food they have eaten or their lack of physical activity. Diet and exercise don't fail! People fail, and they do so when they don't take their health seriously, failing to live in a healthy manner. The advertisements on TV and other media streams say the message clearly: "As I was building my life, cholesterol plaque was building inside of me." The proposal that mysterious forces cause our illnesses engenders fear and is an excellent means to sell products.

One in five adult Americans ingests pharmaceutical anti-depressants daily; Americans also consume 80% of all the manufactured prescription narcotics in the entire world; and many other Americans are self-medicating with alcohol and illegal drugs; therefore, our overall sanity is becoming suspect. Depression and the reason we self-medicate so is perhaps a result of our minds sensing the death occurring within our bodies, which is a difficult position to reconcile when we are surrounded by the comfort and convenience that are the envy of the rest of the world. I also suspect that fibromyalgia and chronic fatigue have this origin and therefore can only be relegated to obscure causes. Both fibromyalgia and chronic fatigue syndrome are poorly delineated by medical testing, and many physicians do not think these maladies even exist at all, but it seems likely that they exist as a constellation of symptoms that are a manifestation of disordered, inauthentic biological lives. Our minds receive information from our bodies, such as fatigue, muscle aches, heat or cold, pressure, and other sensations, information that is only accurate, and therefore valuable, if our physical presence is authentic. Once our physical nature is corrupted, so too is the value of our sensations. It seems a rather simple equation: we cannot expect our feelings to be "good information" if our bodies' sensory organs are dysfunctional. Thus it seems that diagnoses like fibromyalgia and chronic fatigue are in fact an attempt to attach medical nomenclature to the constellation of chronic pains and lethargy, sensations from a body crying out for leadership from the brain, that are typical of these protean conditions. One is left to ponder which is more insane: the disconnect between our lives as we live them and ourselves as biological beings, or the attempt made by physicians to contrive a disease based on a set of vague and disjointed symptoms without recognizing the dysfunctional status of the whole person.

COMING TO OUR SENSES

One fine Sunday morning, I was running through downtown Kaysville on a quaint residential street lined by hundred-year-old trees and small brick houses. This was a cold winter day in Utah, typical in that the trees appeared lifeless from their loss of foliage and the air was extremely dry and crisp. This was one of those glorious Utah mornings where the sky was a brilliant blue and not a cloud was in sight. A foot of snow covered the ground and muffled the ambient sounds. I have experienced numerous moments such as this one, many while jogging on Sunday mornings when half of the local Latter-Day Saints ward was already in church and the other half was preparing for church inside their homes.

I had already run three miles, and so I was sufficiently in my groove and my head was as clear and crisp as the blue sky. At this point I was focused on my breathing and my stride. I was feeling great when I turned the corner at 500 East and 200 South, where I saw a middle-aged woman getting into her car, dressed up in her Sunday best and carrying her LDS scriptures. I would not have paid any attention to her, but her grunts of discomfort broke the silence. She was obviously suffering as she entered her mid-size car, letting out sighs and muffled grumblings of discomfort. I was moved by the moment because I heard distinct, albeit seemingly small, indications of suffering at a moment when I was enjoying an incredible feeling of health. Our ages were not dissimilar, mine and this woman who crammed herself into her car, and as I continued my run past her, I realized that she probably experienced those same uncomfortable feelings every time she got in and out of her car or when she performed other routine activities. I doubted she recognized the suffering evident in her gasps and grunts at a conscious level, however.

When I was overweight, my senses functioned poorly too, and I denied or did not truly "feel" many episodes of discomfort, desensitization and denial, minimizing my discomfort so I could just plug through my daily activities as best I could. I see large people waddle as they walk, and I know they must be chronically uncomfortable but don't recognize the fact. During those times in my life, I did not wish to admit these feelings either.

Many Americans are in fact desensitized and in denial about their small but consistent ill-feelings, such as discomfort when bending over to pick something up, or snug pants, or short breath when chasing children and grandchildren across the yard or when ascending stairs with laundry in hand. Many of my patients complain of generalized aches and pains in their joints, especially in their backs and knees, recognizing these little aches but not really connecting them with their body size. Many also exhibit the excessive sweating and heat intolerance associated with carrying too much fatty insulation, but they don't recognize it as that. A list of such seemingly minor irritations could be quite lengthy, but I believe the reader gets the point that these conditions and sensations are not generally regarded as examples of suffering, possibly because they are incredibly common.

Although the descriptions of suffering I have provided are not thought of as substantial pain or disability, their real importance is increased when we apply a realistic scale of magnitude. Multiply those seemingly innocuous and nuanced forms of pain that are present in the average American's day, felt while accomplishing routine activities, by the hundreds of millions of people in America who suffer from them, and finally multiply that number by the number of days these symptoms are felt (365). I suggest this as a kind of hypothetical pain index, which reveals that these nuanced forms of suffering are not so innocuous.

We could also factor into this hypothetical index the fact that this suffering is not realized solely in the form of corporeal pains but also in mental and economic stress as well. Recently it has been documented that fully two-thirds of bankruptcies in America are a result of medical bills. It has been shown in numerous studies that overweight people earn less money and are less likely to get promotions or new jobs than their thinner peers. Studies also show that absenteeism and loss of productivity is associated with obesity and other chronic medical conditions associated with our most common lifestyle habits. Casualty losses are suffered but the enemy is ill-defined, indeed unnamed. Obviously, my hypothetical daily pain index is impossible to calculate with accuracy, but you get the idea: there is an incredible amount of unnecessary pain in the general population.

My patients report sometimes that there is something missing in their lives — and then they might mention a particular hormone they heard about recently. In the majority of the cases I encounter, there is something missing, all right. Living without cellular corruption has become a distant memory; a life that values physical performance above all else was eradicated long ago. When I tell them that giving up their manufactured food habit would be a good start, they invariably give me a blank stare. I then illustrate what that means. Then as they realize that I mean giving up the foods that they routinely eat, giving them up because they are toxic to their hearts, livers, and brains, almost immediately they appear to have suffered a loss. Obviously, giving up comfort foods can be very uncomfortable.

Physically well people, those who have undergone my thorough examination and have normal x-rays and normal blood tests, nevertheless make something of the same complaint: "I am so tired." I recommend that they should exercise to feel better and have more energy. In response, many often ask for another blood test or a referral to a medical specialist. Exercise is almost unfathomable, and in fact, for most people exercise may not even be on the list labeled "Last Resorts When All Else Fails."

The combination of an unwavering allegiance to manufactured foods with a steadfast resistance to exercise is an accurate predictor of multi-organ failure in the future. To believe otherwise is not only risky but insane, given that our bodies are busy telling us when things are not right, giving us ample time to act. However, once disease sets in, biological authenticity will be all the more foreign and progressively more difficult to reestablish. Although there is always hope, because changing one's lifestyle will change one's biological reality, entry into the medical-industrial complex and the technological promise it offers may be too potent a distraction to take that step. The best approach is to preserve one's senses and one's sanity before it is too late.

However, the difficulty, even for those who are healthy, is duly noted. When I work out less often, I have less interest in working out. This is a sensory issue as my musculoskeletal system becomes far too sensitive to the discomfort associated with a vigorous burst of exercise. Overriding this discomfort for an abstract notion of better health in the future is understandably difficult for those who have yet to make that commitment — it is sometimes difficult for a sworn Naturvore. I find that, unless we physically understand (through direct sensory experiences, as explained in "The Ethic of Self-Preservation") the benefits from exercise, we are not likely to commit to it again and again. Perhaps an ethic of self-preservation will compel some people to put in a performance on days when motivation is lacking. As noted in a previous chapter, I must sometimes ask myself a very fundamental question in order to nullify any reservations to perform: what else would I be doing with this moment of time that is more beneficial to me? My action then becomes the answer to this vitally essential question. My performance during that workout not only affirms my health but also affirms my role in my own health.

Just as we are not typically sensitive to the natural, physical requirements of being well, we also lack sensitivity to the proliferation of disease around us. Most of the people we encounter in a given day are enduring struggles within themselves, a battle with chronic disease. No one seems to be particularly bothered by the fact that disabling conditions such as hypertension, high cholesterol, strokes, heart attacks, and degenerative arthritis that were once reserved for aged adults are increasingly occurring among young adults and children. Kids are being placed on medications once reserved for the aged as a result. In other words, the physical/mental dysfunctions that were once associated with advanced age are becoming the normal experience for all Americans, no matter their age. Tolerance of disease indicates a profound numbness, indeed a senseless state.

Chronic sensory impairment is averted by engaging in a biologically authentic life, wherein the proper function of the body as a whole is maintained. Engaging in regular exercise and eating natural foods grounds us in the most realistic conditions for a truly successful life, a healthy life that is full of genuinely robust feelings. I aspire to being biologically authentic in order not only to feel and interpret most efficiently but also to ensure efficient adaption to any circumstances I encounter in my life. The circumstances of our lives change over the course of our lives, and therefore maintaining adaptability brings additional meaning to the notion of self-improvement. Precise sensory function adds more color and clarity to our lives, but the results are much more important than enhanced aesthetics. We add scope and dimension to our lives through improved sensory function, *and* we can also enjoy an enhanced potential within our lives.

RESILIENCY

As Charles Darwin pointed out, it is not necessarily the strongest or smartest organisms that survive but the organisms most able to adapt. This is in part because strength and/or intellect in one setting may not be strength and/or intellect in another. Life is full of relative qualities rather than absolutes, and therefore our ability to adapt is a primary focus of self-improvement. We must be strong enough and smart enough, but we must also be flexible and able to alter our behaviors and internal physiology to the threats that life throws at us. In times past, humans were threatened more from the outside, but now we are threatened by the chronic ingestion of poisons touted as food or by sinister advertisement messaging which permeates our minds, both capable of making us into people we do not wish to be. Indeed, our biology must find new ways to adapt to these new threats, and the Naturvore identity is my chosen method.

Biological adaptation is an important aspect of a person's resiliency because even our most controlled modern circumstances refuse to be as static and as predictable as we might like. We are exposed to various toxins constantly, most of them products of industrial processes and other human endeavors. Likewise, chronic diseases such as obesity, cancer, and heart disease are very real threats to the modern human organism, and although they have generally been regarded as natural phenomena, they are actually manmade calamities as well. The abundance of manufactured foods, manufactured medicines (one of the largest killers of youth via addictions), and electronic entertainment (which contributes to our tendency to sedentariness) has made modern society deadly. Indeed, far more people die of, and far more suffering is associated with, comfort and convenience than from fighting foreign enemies or gangster killings. We are faced with new existential threats and we must adapt to them or suffer the consequences. Varying degrees of chronic life support are not adequate compensation, nor are they consolation, We can do better for ourselves.

The delivery of healthcare has itself become a threat to the well-being of Americans because it makes people less resilient and more confused about what constitutes health in the first place as well as how to achieve it. Many Americans appear predisposed to fooling themselves into believing they are surviving well within the confines of their chronic diseases, and this quality of tolerance by virtue of modern medicine makes their situation more dire. In these times, threats are much more nuanced and sophisticated, perhaps even counterintuitive, considering the medical industry's purported aims, and therefore our defenses and our adaptations must also be quite sophisticated, at least intellectually. In fact, as part of our adaptation to our modern world, we must monitor our relationship to technology, which inevitably includes questioning and then resisting many of the claims to which we are exposed. In short, for modern humans resiliency — that is the ability to resist disruptions of function and to recover quickly and fully from disease, environmental insults, strenuous conditions or stresses —consists not only of physical defense and biological repair mechanisms but of awareness as well. More than at any other time in our existence, our physical resiliency hinges on an adaptable, resilient and discerning mind. Our reality with regard to fundamental matters, such as our own personal success (Americans believe they are successful sick people), is far more uncertain despite our scientific precision.

I first heard the term "resilient" in 2007, when I served on the Executive Board of the Utah Partnership for Healthy Weight (UPHW), a non-profit organization brought together by Governor Jon Huntsman to curb the obesity rate in Utah. I met with several excellent researchers from Utah State University and found their work on these important matters to be encouraging. The research being done at the university had found glimmers of hope among members of some families in obesity-stricken Hispanic neighborhoods who remained thin or at normal weight despite living in enclaves where obesity is

CHARLES C. HARPE, MD

the norm. These families and individuals were classified as "resilient" because they appeared to resist the normal patterns of poor health in their environment, their neighborhoods. As I sat in various committee meetings and interacted with the researchers, it occurred to me that virtually all Americans live in an obesigenic, sickness-prone environment, where even the youngest people are affected and made sick by a not-so-mysterious force called modern life. Thus, it further occurred to me, resiliency is a virtue to cultivate, and we should all aspire to be resilient humans, capable of repelling threats to our well-being and resisting the cellular corruption that weakens our physical structures.

I have encountered patients from time to time who make themselves more resilient despite the negative influences of our society. For example, Kathryn came to me as a patient on a sunny day in March. She was concerned about her weight gain over the recent holidays. Spring was beginning, and she was looking forward to summer and wanted to be slimmer and healthier by then. Her initial BMI was 35 and her cholesterol 259, LDL 160, HDL 29, HgbA1C of 6.3%, and CRP 8. Her family history contained diabetes but no early heart disease. Her test results were not unusual for an American woman of her age, adding up to a diagnosis of metabolic syndrome or pre-diabetes.

This patient had researched some points to discuss regarding her lab results before she came to my office. Information is of course plentiful via the Internet, but the sheer amount of that information raises the potential for confusion because, without solid experience and education, it is difficult to determine the appropriate context within which to interpret complicated bits of medical information. Patients can also be distracted from sound information by advertisements for less-than-optimal remedies that employ technological appeal and snazzy promises, and of course there is much contradictory information out there as well. One of the sites this patient visited was Crestor.com, where she took a virtual tour of the human arteries to view how plaque potentially builds up in the vascular system. Consequently, the medical-industrial complex had begun to influence our discussion before she set foot in my office. This is often harmful to patients, as it tends to close their minds, trapping them into believing in products for sale without hearing all of the evidence for and against their use.

I choose to empower people to make the correct changes in their lives that will improve their health and their lab numbers, but Kathryn emphatically stated that she needed to have a higher HDL and a lower LDL and came into my office convinced that Crestor would be the best option for her. Although her lab data had more abnormals than simply her cholesterol pattern, it is usually best to engage discussions about new medications one at a time. I define health as being much larger than the symbols, such as HDL, LDL, Hgb A1C and CRP, but within the typical internal medicine office, we must address these endpoints of disease, and so there is little time (or even desire on the part of the patient) to discuss real health as a philosophy of life, a purpose for life or a means to life. As many patients do, she entered my office focused on her cholesterol, but as I completed a thorough review of her body systems, I observed that she had multiple complaints not attributable to cholesterol deposits. She told me she was not sleeping well, was tired more than she thought she should be, and had unusual migratory aches and pains. Unfortunately, these are not uncommon complaints to a primary care physician. Within our population it is exceedingly common to find people sensing a slow death, but unsure of what they are feeling: the loss of vital energy.

Nevertheless, we proceeded to tackle Kathryn's cholesterol readings since they seemed to be what prompted the visit (compliments to the advertiser for Crestor). I told her she could exercise and increase her consumption of omega-3 fats to boost her HDL while eating less saturated fat to lower her LDL cholesterol. I encouraged her to shed 10-15 pounds, which is the best way to achieve a better cholesterol reading and also tends to improve one's life in general. In fact, increasing HDL through medication use has not been shown to improve people's mortality risk, but exercise-induced HDL increases do seem to protect people from heart disease. For several years, possibly decades, medication manufacturers and subsequently physicians (I am included) touted that higher HDL levels from medications would potentially save lives, but this was, dangerously, extrapolated from observational research studies that found an association with higher HDL levels and fewer cardiovascular events in the population of "healthy people." Follow-up research indicates that increased medication-induced HDL levels did nothing to spare cardiac deaths, however. Unfortunately, the original information sponsored (spoon-fed) by the pharmaceutical company became a mantra for most physicians before all of the results were in. (Still worse, at the time of this writing there are additional concerns for safety and the advisability of prescribing statin-type cholesterol drugs to middle-aged women at all.)

As we discussed these issues, Kathryn explained to me, as many people often do, that she was under a lot of stress at work and could not possibly engage in routine exercise in the near future. Furthermore, she said that she could not change her lousy diet because she ate with the corporate executives, meals catered by fine restaurants. I told her that a better diet and exercise would give her much more health than just better cholesterol, that she would also benefit by improving the organ dysfunction relating to her metabolic syndrome and reduce her risks for breast cancer, colon cancer (obesity is the number one risk factor for colon cancer), hypertension, diabetes, osteoporosis, strokes, and heart attacks. In other words, I told her, she could become resilient to all of the common afflictions of American women. I pointed out that, if there is any true magic in this world, it is diet and exercise. Only 20 or so minutes of exercise most days of the week, when combined with a healthy diet, is all that is necessary to make a large difference in one's life. I suggested to her to run just one mile per day for 5 days a week, or to run 5 miles per week no matter how she could work it in.

The interactions I have described thus far are not at all unusual. We can see that Kathryn was "broken" as an organism, not being true to herself and the nature of her life because she did not consider her routine actions with regard to her lifestyle as very important to her health. Her excuse, also not unusual, was that she was busy serving other needs, particularly those aligned to the Industrial Order, and was too busy to serve her own needs for health. She came to me demonstrating laboratory evidence of health issues, but she was in fact showing signs that are more substantial to the assessment of health: not sleeping well, experiencing excessive fatigue, and being overweight. My patient was living a less authentic life as an organism. She was not tending to her rudimentary biological needs, those that allow her to feel good in her own skin. She had lost touch with her essence, her exquisite anatomy, and the biochemical functions that create her life.

I enjoy counseling my patients because I believe in the message of unity in our biological energies, those arising from the mind and the body, more than I believe in the PDR (the *Physicians' Desk Reference*, a product of the pharmaceutical companies), because the latter is for sickness and I am interested in my patients' health. Kathryn had been thoroughly prepared (manipulated) for our patient-physician encounter by the Crestor TV commercials and the arterytour.com web-based tutorial on atherosclerosis, and so the discussion became lively, as one would expect from an engaging administrative secretary to a corporate president, a smart person who arrived prepared with an agenda. Kathryn is in fact a perfect example of the Industrial Order's empowered patient, but, unfortunately, such patients are merely empowered consumers of the pharmaceutical companies.

The easiest path is the expected path for a physician: to write a prescription. I could have given this patient a prescription for Crestor and she would have been satisfied, but I would not have addressed her real problems and I would have stolen her autonomous energy (she definitely had some vitality in her). She had yet to demonstrate to me that she could not perform what I asked of her, improve her cholesterol by diet and exercise. Furthermore, I made her aware that her problems of insomnia and excess stress would get better with exercise and some self-care. Women tend to be care-givers and are often very busy with jobs and families, therefore have many people to tend to, but I remind them that their boss, their kids, husband, even their doctor will not care for them. I told her she had to take care of herself because it was her responsibility and no one else would do it.

She acceded to my request, and we parted company with a sound relationship. I sensed she felt somewhat empty-handed while departing from my office, however, largely because she had done a good deal of research before she came to me, all of it leading to a magic pill as the answer to all her problems. Fortunately, I detected an open mind as well as one prone to solving problems with research. She was willing to reexamine her lifestyle so she could make the changes necessary to avoid the devastating consequences not only of high cholesterol but also of diabetes, heart attacks, cancer, and strokes.

Kathryn returned to my office three months after our initial visit to report "progress" in her efforts to eat more natural foods and increase her exercise. This motivated patient began running as I had encouraged her to do, although not quite to the level I recommended. She was still doing very well by running almost one mile at a time most days of the week. As a result of some simple changes to her habits, she felt more energy and vitality, and I could sense more peace emanating from within her. I asked her if she noticed that she was more able to actually feel the health of her body, and she affirmed that she did in fact feel healthier. As she described some of her activities, I noted that her senses were becoming alive again, just as her body had more life. She also told me that she felt more confidence, even in the stressful and competitive atmosphere at work.

We build our resiliency through biological authenticity and by practicing healthy behaviors. We also discover the knowledge most important to our health — our biological potential. This means we can actually *feel* our health. How else would an organism be able to adapt if it were not living with finely-tuned senses? How would nature tell me that I was eating the wrong foods or that I had remained busily engaged in activities that were less conducive to optimal health? If people did not change in physical appearances, how would they know that an incredible amount of damage is being done to their bodies at the microscopic level? Medical tests do not provide the feedback that will make us more resilient people. Any change of health fate through one's actions is always superior to improving biochemical tests.

By the time Kathryn had followed-up in my office, she had lost 5 pounds and was sleeping better, and therefore she had much more energy in general. Lack of sleep is often viewed as a problem in and of itself (and Lunesta and Ambien commercials inform us that insomnia is a medical condition and therefore worthy of their industrial treatment), but in actuality, it is the lack of a biologically sound day that leads to most insomnia. All biologically authentic creatures sleep well, period. It is fundamental to life as mammals. Therefore, the inability to do something so basic means we are somehow living an inauthentic life. The best way to sleep is to focus on reducing the effects of the Industrial Order in our lives, adjusting the demands of society upon us, and remaining authentic organisms through a natural diet and exercise.

Although, she had yet to reach her ideal weight before summer, Kathryn obviously felt better in general, and I could tell she was developing good behavior patterns that would serve her for life. That three months would not have been wasted in the larger scheme of her life even if she had not lost any weight, however, because she was making important changes to her life. In fact, it takes a bit of time for people to get in a rhythm of practiced health, and allowing a patient six months to a year to develop good self-care health habits is not unusual.

Approximately seven months later, just before the holidays, Kathryn had lost another 8 pounds. Now she was on fire with passion for this new way of life. She visited to pick my brain about my exercise routines and to share her dietary changes with me. She told me she had more energy for her children and enjoyed time outdoors with them. She wanted to continue her efforts and did not want another cholesterol test or a medication prescription at that time. One and a half years later, she returned to the office, 20 pounds lighter and down two dress sizes. We repeated her blood work, and indeed her cholesterol profile had improved, her pulse was slower, and she reported that she felt much better in general. Her medical goals were being met as defined by the medical societies, but, more important, she was gaining in more human terms. She had increased her resiliency by her own power and physical resources. That is not to say that along the way she did not have many weeks where the weight was not falling off as much as she desired, but she stuck with it as a resilient person does. She did not need science to make a numbers game of her life. She was living the extraordinary life of a person invested in her own presence, her personal health.

Kathryn created her own health fate, one that she was not aware she had within her before she attempted to obtain it, when she managed to escape her well-formed Industrial Order identity and paid no heed to the ads assuring her she needed to take a pill. In cases such as this patient's, in fact, I believe it is inappropriate to prescribe medicine, even though most physicians in the medical community would have done so. Luckily, however, some patients become resilient to a culture that creates more sickness than health. Kathryn became invested in her life when she realized that, if she was not busy working at being a biologically healthy person, then what was the point of her other life works? Samuel Butler once said, "Every man's work, whether it be literature or music or pictures or architecture or anything else, is always a portrait of himself." This applies to health as much as to any other aspect of life. My patient began to view her life and her health as one entity and thereby became an authentic biological being and reduced her chances of diabetes, cancer, heart disease, strokes, Alzheimer's dementia, sleep apnea, hypertension, kidney failure, atrial-fibrillation, debilitating arthritis, and the other dreaded chronic diseases that are taking an increasing toll on our society. This is true resiliency and cannot be purchased from the medical community.

THE SEAMLESS NATURAL EXISTENCE

People who come to me for medical treatment are frequently "broken," their life energy incoherent, long before their abnormal medical scans and test results. They are suffering from a severe mind-body disconnect, which also resembles the disconnect people have from the natural world, wherein they are deemed biological organisms but do not actually live in a manner that reflects that reality. If we understand how energy is moved and utilized by our bodies, responding to a life of purpose, it is easier to understand the seamless nature of our existence and how best to use this quality to secure optimal health.

I understand very well the power of industry to thus dichotomize us and reduce us to some input for technological medicine. I was once a physical form barely resembling a biological being let alone a seamless being, wherein my body and mind are a single entity. When my mind was held with higher regard than the other parts of my body, I valued things in life a bit differently. Science, by intellectual means, easily squelches our natural spirit and function via reductionist methods, diminishing our natural abilities until they barely exist at all, and in the process makes us less healthy. This is not the fate we need to accept, however, as I have argued throughout this book: when we value wellness above all other things, we tend to be well. Self-preservation is not an intellectual endeavor but rather a value judgment.

As I accumulated weight around my abdomen and throughout the rest of my body, it was visible on the outside and for all intents and purpose was an obvious sign of dysfunction, yet the outside fat was seamlessly creating dysfunction inside me or the reverse, no matter how one looks at the situation. My mind was feeding my body junk, not concerned about the cellular milieus within, so my body also took a hiatus from my health. If the mind is distracted from health, of course the body is as well. Many people appear to understand that what they put into their mouths can translate seamlessly to the dying, gangrenous toes associated with diabetes or even multi-organ failure. Indeed, we must be sufficiently self-aware as biological beings so that we are able to discern for ourselves when we are sick and when we are well rather than relying on the Industrial Order to parse our health into a set of seemingly disconnected numbers, further breaking up our seamless biological nature. Our seamless nature with respect to energy transfers within authentic biological tissues elevates us above a life described by science; we are instead busy harnessing the force of life.

Consider our bodies as boats that, with time and use, will develop a few holes. Obviously, a better captain has fewer holes in his or her boat, and staying afloat in a lesser captain's boat will require some intervention to keep it going. The expert application of very sophisticated patches ensues, the most technologically advanced patch kits available applied by the most expert patch specialist. The repeated application of these complex patches keeps water out of the boat, but this can be viewed as of little importance when we realize that the boat is headed for a large waterfall. Our focus on the placement of the patches, even if they are very good ones, can leave the entire context of the boat unattended to, and the fate of the boat is much worse, despite the patchwork. In short, we must be good captains who know better than the patchers what is right and wrong to do in our lives. As seasoned and practiced captains of our bodies, we can see a more complete reality of what is being proposed and what is being done to us by those who are supposed to be interested in our health. However, as noted, the scheme of modern medicine simply reflects the brokenness, the disconnected state, that people have already established for themselves before they enter the medical plan. Americans are typically minds (the captain) working against the interests of their bodies (the boat) by denying the greater context of their health (the body of water in which the boat sails) and relying on experts to fix one hole after another.

The tension we sustain between the Natural Order and the Industrial Order looms large within our lives, but it is perhaps most palpable in the setting of healthcare precisely because the Industrial Order version of health is anything but. Indeed, our children, the future of our country, are becoming entrenched in the Industrial Order even more forcefully than previous generations. The very biological resiliency of everyone is being challenged. Virtually every child, even the poorest through Medicaid, has access to routine blood tests and potent medications or procedures, but increasingly children in the US have little or no access to meaningful vigorous physical education and none have an optimal diet at school. Thus, even our children's health is being measured by their medical test numbers rather than their physical performance, and they are being treated as inputs to the medical-industrial complex at ever younger ages as a result.

The much revered academies of pediatricians and endocrinologists (specialists of hormones and cholesterol) are recommending that all children should be screened for high cholesterol, and screened as young as eight years of age. This is so that "interventions" can take place. Because every child should be engaged in healthy activities, not just those that are measured in the doctor's office, then what "interventions" are they suggesting? Both academies recommend that children be placed on potent cholesterol medications such as Lipitor if they have high cholesterol. In fact, Pfizer Pharmaceutical Company is making grape-flavored chewable Lipitor tablets (Lipitor candy) for children, although they are currently only available in the United Kingdom. The ethic of self-preservation is being replaced for even our children by an artificially contrived, less autonomous human existence.

Children are being encouraged to rely on the medical experts created from the Industrial Order, to become fully invested in fancy technology, rather than being taught to understand that health is derived from within themselves by what they do. The fact that gastric bypass is available to adolescents, as are weight-loss medications, is further proof of the insidiousness of the Industrial Order. The future of our country would be better served by implementing physical education, eating natural foods, and combining these with a resounding respect for the Natural Order. Children should place value on their overall health as the best guarantee to securing a better life, the ultimate condition of self-respect. Here is a lesson for our children, one that would serve them well the rest of their lives: Health is realized through biological authenticity, and resiliency in our lives is built by intrinsic means, specifically through a unity of mind, body, and spirit. Each of us can opt out of the cultural madness, but first we must see it. The ability to discern that which is artificial, including an artificially proposed purpose for life or consumerism, from something that is natural is a necessary survival skill. Obviously, internalizing the Industrial Order invites personal chaos because our most authentic forms come through the Natural Order.

Americans in excellent health already stand out like resilient islands in a sea of chronic disease, but obviously they will be more rare in the future if this nation stays on the path it is on now. The statistics concerning the health of our nation are expected to worsen because the medical community is powerless to stop cellular corruption and the corruption of our minds. Only by unifying our bodies and minds into a seamless entity do we create the most resilient forms of ourselves, a self worth preserving, our most important form of capital. Without self-preservation who or what are we?

Our purpose should mirror the realities of our seamless natural existence rather than an artificial existence. Showing up for a vigorous workout and eating natural, whole foods signify a completely different set of intentions for life than do sitting for hours passively entertained in front of the TV, eating manufactured foods, and ingesting manufactured medicines. We can be vital or we can attempt compensation for the loss of a vital life. The key to understanding our need to engage with the Natural Order and to stand defiantly against the Industrial Order is then simply stated: resiliency always begins from within us, not from some applied technological fix after the fact, after we have decayed sufficiently to enter the medical-industrial complex utterly. A healthy life becomes much easier when we realize that our internal physiological circumstances are as important, if not more important, to us than our external circumstances, especially those of the Industrial Order.

The resiliency we sustain throughout our lives will surely change the course of those lives, just as a lack of resiliency would. By also being resilient to the Industrial Order, we can preserve our biological authenticity. However, as noted previously, our intellect is not necessarily the best tool for establishing resiliency in this day. Indeed, it could be argued that resiliency requires a reawakening of primordial wisdom that becomes a new adaptation, inasmuch as we need to bring that wisdom to the level of consciousness.

RENEWAL OF PRIMORDIAL WISDOM

At forty-two years of age, I went from a 36-inch waist to a 29-inch waist and from getting winded walking a single flight of stairs to running a half marathon in one hour and forty-five minutes. These fantastic changes did not arise from medical science or even my intellectual investigation into my health. The physical and mental transformation arose as if from within me, from some primordial center beyond the reach of intellection but that I would nevertheless label wisdom. This wisdom came from my body and was not formulated anew but uncovered.

This transformative period had all of the markings of a grand intellectual design of a trained physician, and indeed I have presented many examples of sophisticated scientific understandings in this book, but the original discovery was not intellectual. I simply use scientific explanations to fight fire with fire, so to speak, resorting to the methods of the medical-industrial complex to undo their influence and control over our affairs. The changes I made to my life were in fact of a primordial origin, the optimally healthy life available to all creatures.

I am convinced that the cells of my body were leading me through the fantastic changes within me and that my mental faculties were simply following along. I was translated, in a very physical sense, to a more secure, more authentic human form at a cellular level before I changed, in an intellectual sense, for physiological synergy and synchronicity (they create man's intellect, not the reverse) are far more potent than intellection. I simply began to live in line with the biological systems that I was born with, and, by establishing a more authentic physical form, I serendipitously encountered my real identity: the Naturvore.

My body took on the physical form, and hence the true identity, of the Naturvore long before my conscious mind was even convinced my actual identity had changed at all, let alone invented a name for that change. In other words, I became more real in physical form before my mind was made more real. The primordial wisdom that led me to become a Naturvore was not a product of the outer layer of my cerebral cortex, where our higher intellectual centers are found, but from within my body. After sufficient change was sustained in my overall internal milieu, my mind was also changed and eventually improved enough to discern a new reality. Viewed slightly differently, the intellectual parts of my mind were forced to make sense of this new personal identity that it had no previous concept for, and even less input into making, and it was at this point that this new personal, physical, and mental identity deserved a name.

Medical-industrial scientists will argue my story is simply anecdotal and that the changes I have made in my life should not be even considered as a means to a healthy person much less a healthy population. My life has not been scientifically studied, and therefore, according to the reductionist rules of the medical industry, my health habits hold little value to the medical establishment. However, as I have argued throughout this book, we can sense when we are well, just as we can sense when we are not, unless we are senseless beings. Consequently, I have offered up my own sense of physical well-being, my own sense of my biological authenticity, as proof that we do indeed have a choice when it comes to what constitutes our health.

My body became resilient and authentic (as in my real human body) over a few months of doing very unintellectual things such as eating less food, eating a natural diet, and performing exercise. Accompanying these physical changes was an increasing sense of a new personal value, empowerment, and a different pattern of thinking. I now view my health not as a product of intellectual propriety or innovations attributed to the scientific method but as a product of my personal values. Because I began to change so quickly and seemingly with ease, my body, my physique, in effect surprised my mind.

Our intrinsic worth, our personal worth, is derived from our minds' relationship to our bodies. Extrinsic worth, instrumental worth, or even extrinsic assessments by others may be potentially gratifying on some level but generally add little value to our lives. Placing more value on the extrinsic world (society and material things) rather than the intrinsic world (our bodies) means we are allowing the artificial world to provide our personal value rather than the natural world. Thus the term Naturvore signifies not only a new physical form but also a new respect for the natural world and the intrinsic value nature provides via an authentic biological existence. My Naturvore identity also represents a transition to a more discerning mind, one that realizes that my real nature was not an industrial innovation or a product of the scientific method.

It is in fact nothing short of ironic that we make health and the achievement of a better physical life so intellectually difficult when true health is utterly straightforward and distinctly unintellectual. True health is merely a function of uncovering what has been within us all along. The physical reality I have now has been stored in me since birth but could not be realized until I removed the excess lard and industrial-made ooze encasing my organs. I had to be made more physically authentic before I could be made more mentally authentic. We are physically human before we are mentally human, the mind only coming to life as an entity when the body it is associated with is alive. Our complete physical forms are our real identities and they reflect how we view life overall, especially our own.

When a bald eagle chick leaps out of the nest for the first time in its life, there is absolutely nothing rational or intellectual about that action or the fact that it believes it can fly. The eagle has the physique of an eagle and that is what tells it that it can fly. The wisdom of the eagle is held within its entire form, not the eagle brain. As with an eagle, although we would perhaps like to believe differently, our wisdom is also stored within our physical form. The eagle begins adult life with the most irrational behavior, leaping from the nest high above the ground, but possesses an internal wisdom. Indeed, the bald eagle chick flies by virtue of a wisdom that is intrinsic to its biology — it simply does what it is supposed to do. Likewise, analytical knowledge, constructs of the latest evolutionary attributes of the proud gray matter of the human cerebral cortex, provides us little useful information for actually creating our lives. We need very little intellect to survive very well in life (I suspect the reader knows someone who proves this point). Instead, most of our life is generated and maintained from areas in our brains far removed from the intellectual centers, from the subconscious centers that actually maintain the workings of our bodies. Information essential to achieving a healthy, happy life lies in the form of instincts within the circuits of the most primitive sections of our brains and neuron-genetic imprinting that affects all aspects of our lives with little alteration from intellect.

The human intellect is mesmerized by the Industrial Order, which is after all the epitome of human intellectual capacity. We tend to invest the majority of our lives and resources in the Industrial Order, but, as with our intellects, it does not tend to create the most authentic human forms of life, which brings us back to the notion of true wisdom, as in the wisdom we intuitively know is real because we can feel it. I suspect that a body that is more authentic and wiser about itself also creates the conditions for more rational thoughts. In other words, by improving our bodies, we improve the contents of our minds.

Body neglect is brain neglect. Dementias and strokes are definitive losses for the mind. Clearly, as the evidence is only now being reported, both of these conditions are directly attributable to the ongoing cellular corruption that initially becomes more obvious in other parts of our bodies, such as obesity, diabetes and heart disease. So, for those people who live within the dichotomy of mind and body: beware. Emphatically, a more authentic body yields a more authentic identity, and hence a more authentically human mind; and a more authentic body, a more resilient body, automatically creates a more resilient mind. To be an optimal human, one must act in accordance with authentic human nature.

The shape of our physical identity gives shape to our mental identity more than the reverse. Or, said another way: a more authentically-shaped human form provides for a more authentic human identity. The proof is in our interoceptive neuronal networks. Neurologist Dan Siegel, MD, points out quite compellingly, "The neural net processors around our internal organs directly influence our reasoning. The hormonal state of our bodies and the tension in our muscles, limbs, torso, and face, each contribute directly to how our interior world shapes our feelings" (*The Mindful Brain*, 2007). Interoception can in fact be made keener by consistently aligning our lives to our biological authenticity. Conversely, if we use our bodies less, we will be less connected with them, and indeed interoception, like other subconscious sensory networks, can become corrupted from its original form, function, and purpose.

During the first forty years of my life, I was complicit with the Industrial Order and measured my success by the things I acquired within it. My physical health suffered as I was pursuing values that were deemed successful in society, values very different from the intrinsic biological values by which I now live. I was thinking of a larger 401k, a faster car, and a bigger house, while my organs were becoming increasingly corrupted, losing not only vitality but their original purpose. I had lost touch with my internal nature, the biorhythms, the ecosystems, and the universe within me.

Living according to the Industrial Order feels so very rational, a desire for freedom, as represented by our technological innovations, pursued with supreme purpose. But as I achieved a more authentic body I began to realize that my real freedom could be lost to the very lifestyle that was supposed to symbolize ultimate freedom. Indeed, I discovered that health provides the most freedom in life, and the lack of health is a significant constraint. It is unfortunate that freedom in America is often mistaken for the ability to choose one's oxygen company, doctor, hospital, face mask to breathe, CT scan machine, sugar testing kit, and brand of drugs.

Often mentioned in marketing circles is the "pulse of the consumer," a measurement industry uses to target pitches for their merchandise, whether it is cars, radios, hamburgers, or medical equipment. The metaphor is telling. The pulse is, of course, metaphorical only and has to do with consumers' minds, not their actual arteries. So, even our pulse, the measurement of our very heartbeat, is an abstraction, one used to determine what can best be sold to us. Moreover, consumers' needs and desires are not so much being measured as created by the products themselves. The Industrial Order's existential concern is to turn our desires into needs — and modern medicine is a prime example of this.

So, our people are being told what they medically "need," when realistically what they want is complicity with their disconnected state from nature, that which is being sold to them. Our proverbial pulse is being consistently taken and our desires fulfilled, but healthcare has less to do with a pulsation emanating from a heart than customer care and delivery of a pill or procedure. If we pay heed to our actual pulses as they represent our actual heartbeats, our actual health, then we would pursue something else entirely, a path revealed by primordial wisdom: a vital life invested in better food and strenuous activities.

Picasso said that art is the lie that shows us the truth, which I would modify relative to the very dangerous situation that is our modern life relative to our health: science is the truth that conceals the lie. Through science and the evidenced-based medical paradigm, Americans are increasingly compelled to believe that it is normal to require several specialists to sustain each one of our lives. Western physicians, especially the specialists, may be well-reasoned, intellectual professionals who can reduce our problems to a single number and a pill to "fix" it, but in fact they lack concern for, and indeed knowledge of, us as larger biological organisms.

My patients often complain that specialists do not listen to them or acknowledge aspects of their lives that to them seem important to the disease in question. I remind them that specialists have chosen to limit their scope of practice and that is why they are called specialists. They choose not to engage with the larger concerns of a person's life because that will make their work more difficult and less satisfying as a specialist. Courtesy of powerful reductionism, the specialist has just spent several years in training to hone his focus to an increasingly narrowed interest and thereby to become an expert in a subject of very limited scope. This reductive move is definitely misleading if we consider it as being good. We are favoring the small view of life over a larger view, while losing broader skills and talents from our society. Such a narrowed view conceals glaring defects, in that medical specialists are not certified in health but rather in disease. A very important point to discern if we are intent on creating health.

For example, Americans tend to assume that, as our stroke centers become more numerous, we are truly living in times of great stroke knowledge. It occurs to very few people that we construct so many medical centers not because our culture understands a great deal about health but because our culture understands so little about it. The Industrial Order builds more and better stroke centers, first and foremost, because that is what the Industrial Order knows how to do (that and making a profit, of course). Primordial wisdom maintains that it is best not to have strokes in the first place, but this wisdom is supplanted by our intellectual (and financial) desire to build stroke centers.

Stroke-care technology does not bring us closer to understanding people, and therefore that technology fails to prevent strokes. Indeed, as noted previously, stroke centers do a very good job of assimilating strokes into the scheme of a routine American life, yet another consolation provided by the Industrial Order, another scheme of chronic disability as opposed to sound bodies and minds. But we are only consoled by these examples of scientific innovation if the real truth remains concealed from us: our population must sustain a certain amount of strokes per year to justify the operational expense of these fantastic units. A substantial flow of people are now required to sustain the application of resources priming the helicopters, CT scan machines, clot-busting drugs that cost two thousand dollars a dose, specialized personnel, and various miscellaneous pieces of equipment poised to rescue us in an instant. Centers that care for strokes may bestow a sense of comfort, may appear intellectually justified, but only to the least discerning minds. Indeed, the most discerning mind knows that making more stroke centers means more strokes, that the resources and intellectual interests of the Industrial Order do not lie with people but with the strokes that afflict them. Investments made in stroke care are a very narrow concern of life.

The primordial wisdom, which reveals that people are supposed to be well, which grows from living as authentic biological beings, living as humans have evolved to live, the wisdom inherent in biological authenticity, is being replaced by healthcare knowledge and evidenced-based medicine. We are losing the ability to tap the biological wisdom inherent in our form and function as authentic biological entities, replacing it with a blind belief in technology. The primordial wisdom within us organizes our tissues in such a way as to create an optimal life by prompting us to do things that are ancient and simple. This wisdom that guides my choices, avoiding disease provocation, is the very same biological wisdom that trains my body to resist disease. We are seamless in existential matters.

Although we generally do not view a vigorous swim, bike ride, or eating a fruit as a life-altering event, they really are, and that is what our primordial wisdom is trying to tell us: to drastically alter our biology, our minds, bodies, our complete fate in life, all we need to do is place one foot in front of the other and run, go to the gym, or visit the local farmer's market. I engage in positive actions because the alternative is, frankly, negative. There is no middle ground in biology. We are building biological vitality or simply atrophying. We are either cultivating resiliency or cultivating disease.

The Naturvore identity is my refuge within the Industrial Order that impinges upon my primordial wisdom at every turn, but enactment of that identity is also self-reinforcing. After a workout, I am put right with the universe, and I know that what I am feeling is more authentic, more sincerely related to what I am supposed to feel in this world. Interestingly, the same physical activity that strengthens my body also provides the wisdom for my food choices because I am nourishing something that holds more value to me. From the strength of my Naturvore identity, I can easily choose whole foods rather than factory-built foods. I can choose to run today, to work out tomorrow, to live completely every day.

Again, the Naturvore identity is self-reinforcing. Food and activity choices that represent a more authentic human living experience instruct our bodies in how to live better, and this positive energy indeed transforms our minds to a more healthy state even as we make our way daily through the Industrial Order world. As a result, the choice made initially to defy the Industrial Order becomes our response as a matter of course — naturally, if you will. It becomes, in fact, liberating to have eating patterns that are both intentional, a function of choice, and less an intellectual game wherein micronutrients must be managed; and by improving strength, flexibility, resiliency, and authenticity in the body, a mirror reaction occurs in the mind. A more biologically authentic body not only reflects an authentic mind but also creates an authentic mind.

In fact, I have become more aware of particular fears and desires that may corrupt my physical attributes. When I feel unsure about the future, I can always refer back to my body to provide context for my concerns. For instance, I make career and work decisions based upon my health now (a better guaranteed future). My mind will typically prod me to make more money, doing whatever is readily available, with little concern for the rest of me. Therefore, I have found it wiser to channel my proverbial brightest ideas through the filter of wisdom innate in a sound body, to make sure my decisions will be good for me as a whole.

As a result of my Naturvore identity, my relationship with technology has changed for the better. I more clearly see its perils, its enticements, and its ability to transform our lives negatively, to make us weaker and less resilient. These days my circumstances are negotiated from a position of strength and health rather than allowing them to be dictated from third parties and professionals. I consistently manage the technology to which I am exposed, every instance being not only a test of my relationship to the Natural Order but also a process helping me to understand the influences of the Industrial Order by virtue of a healthy medium. The Natural Order within me is the master of the Industrial Order around me, not the reverse.

I consider moments of insight, particularly those pertaining to the rediscovery of our primordial wisdom, as means to overcoming the Industrial Order, as proof of a sublime life through the art of self-preservation. In my travels since I have chosen to live as a Naturvore, I have had numerous moments of such transcendence, but one example will suffice. In April of 2008, I was waiting in the Atlanta airport to fly into Salt Lake City, and I noted the obvious: the airport terminal was entirely artificial, 100% built by the will and technological wizardry of humans, and then I noted that everyone around me was eating artificial foods. A couple of years before this moment, I too would have been engaged in the cellular corruption that I was now witnessing.

During this layover, I had the opportunity to eat my second meal of the day. I had brought with me four ounces of lean pork seasoned with turmeric, black pepper, chipotle pepper, garlic, and olive oil. My carbohydrates sources were two apples, a Jonagold and a Cameo. A strong feeling of familiarity came over me in this moment because I was eating one of my usual food staples, but, because I was eating it in a very unusual milieu, the contrast struck me as poignant. The comfort of my home-cooked pork and two favorite apples was most welcome, and indeed I had never traveled to the airport with my food before, but the Naturvore identity made this seem not in the least out of the ordinary. In fact, I found myself feeling quite healthy by comparison to my peers on this journey.

I could sense the steady infusion of optimal nutrients coursing through my body as I engaged in mindful, meditative contemplation while I ate. This transcendental meal made me feel happy and content, as I imagine animals must feel when they eat the foods they are supposed to eat. I, like a contented squirrel, sat in the Atlanta airport terminal focused on a perfect cellular moment because I was eating exactly what I was supposed to be eating; and this feeling was perhaps all the more poignant for, unlike the squirrel who does not have to make such decisions, I was not participating in the artificial comfort of French fries or a greasy burger constructed by food scientists in a far-away corporate food laboratory, as those around me were. In fact, the majority of patrons in this food mall around me were eating two to three times the calories they needed to meet the demands of their travels. I witnessed people consuming extraordinarily unnatural foods that cause total body inflammation and many of the dreadful industrial diseases. From my perspective, there did not appear to be any thought regarding the biological moment they were having, nor did the people around me appear particularly healthy. I doubted any of them felt healthy.

The air was filled with artificial scents, especially the scents of hot grease and fried carbohydrates. Everyone appeared to be in denial as to the harm they were self-inflicting. Of course, when pressed as to why one would engage in such unhealthy behavior, I am sure many people would answer that this is not the proper time or place to entertain healthy thoughts, much less the ethic of self-preservation. Pressing issues, such as catching a plane, or perhaps merely the rationalization that just this one meal would not hurt them, suffice to override any notion of self-preservation they might have. Despite this most artificial environment and the onslaught of smells that would have once had me salivating, however, I remained aligned with nature, natural food constituents coursing through my body to supply my cells with optimal nutrients.

I also noted that the advertisements, the bright signs and the multi-colored establishments that surrounded me, did not speak to me as they once had, because the fast foods or other reconstituted food substitutes such as the "cinnabon" did not appeal to me as they once had. I was savoring a meal that came directly from the earth, and I felt very comfortable with the fact that I could "opt out" of the excess calories, fat, and preservatives surrounding me. Instead of hurting my body, I was replenishing my natural fabric with natural products while seated in the Atlanta airport waiting for a plane to take me to my family. I noted with satisfaction that, even in the Atlanta airport, I had resolutely managed my relationship with technology and with the Industrial Order. I used technology for travel but did not consume artificial foods made through technology.

The lesson of these transcendent moments is clear: the Naturvore identity is a coherent belief system through which my cellular integrity makes me resilient not only to industrial-made substances but to industrial-made messages as well. My wish is to remain a resilient, discerning creature, one that can sense the boundary between the Natural Order and the Industrial Order and to be able to operate within the latter as a product of the former. This is teleologically satisfying for this simple organism. We can all be powerful governors of our health and need not succumb to the artificial proposition of life, wherein our bodies are sustained by the products of industry and our hope for the future lies only with man's innovations. Moreover, it means we can opt out of the Industrial Order version while still living within it merely by making choices based on the Naturvore philosophy.

Stated another way, living as a Naturvore allows us to opt out of the cultural insanity of manufactured food and industrial medicine while renewing our faith in the wisdom inherent in our biological beings, our primordial wisdom that indicates what a vital human really is and how to achieve optimal health. The Naturvore identity is new and yet connects us to our primordial past, the most fundamental means to securing a healthy life; therefore this identity represents a new human adaptation to improve our lives in modern times. Instead of relying on man's whimsical inventions, especially those that become enmeshed within our physical fabric, we can revere the seamless natural life that translates into better organs, better bodies, and, even further up the complexity ladder, a better society. I am a better person for my family and my neighbors when I am healthy and seeking to improve myself further, which is contagious, and so it is possible to imagine just such a change. That is, the key to building a better society is to build better biological identities that actually represent biologically authentic people. Nature has entrusted us with this gift of life, and it is up to us to be true to that gift.

APPENDIX

BOOKS

Aday, Lu Ann. "Reinventing Public Health: Policies and Practices for a Healthy Nation." San Francisco, Ca, Jossey-Bass, 2005.

Brownell, Kelly D. PhD. "Food Fight: The Inside Story of the Food Industry, America's Obesity Crisis and What We Can Do About It." New York, NY, McGraw Hill, 2004.

Buettner, Dan. "The Blue Zones: Lessons For Living Longer From The People Who've Lived the Longest." Washington, D.C, National Geographic Society, 2008.

Conrad, Peter. "The Medicalization of Society: On the Transformation of Human Conditions into Treatable Disorders." Baltimore, MD, The Johns Hopkins University Press, 2007.

Geyman, John M.D. "The Corrosion of Medicine: Can the Profession Reclaim its Moral Legacy?" Monroe, ME, Common Courage Press, 2008.

Hadler, Nortin M. M.D. "Rethinking Aging: Growing Old and Living Well in an Overtreated Society." Chapel Hill, NC, The University of North Carolina Press, 2011.

Hadler, Nortin M. M.D. "Worried Sick: A Prescription for Health in an Overtreated America." Chapel Hill, NC, The University of North Carolina Press, 2008.

Harrington, Anne. "The Cure Within: A History of Mind-Body Medicine." New York, NY, W.W. Norton and Company Inc., 2008.

Pollan, Michael. "The Omnivore's Dilemma: A Natural History of Four Meals." New York, NY, The Penguin Press, 2006.

Rifkin, Erik PhD; Bouwer, Edward PhD. "The Illusion of Certainty: Health Benefits and Risks." New York, NY, Springer Science, 2007.

Roberts, Paul. "The End of Food" Boston, MA; New York, NY, Houghton Mifflin Company, 2008.

Siegel, Daniel J. "The Mindful Brain: Reflection and Attunement in the Cultivation of Well-Being." New York, NY, W.W. Norton and Company Inc., 2007.

PERIODICALS

Evidence that protein requirements have been significantly underestimated.Elango R, Humayun MA, Ball RO, Pencharz PB. Curr Opin Clin Nutr Metab Care. 2010 Jan;13(1):52-7.

Underappreciated Problem; Annals of Internal Medicine, April 3, 2012, vol.156, no.7

Punishing Health Care Fraud -- Is the GSK settlement Sufficient?; NEJM; Outterson, J.D.; Sept. 20, 2012; 367;12

Overdiagnosis in Breast Cancer Screening: Time to Tackle an Underappreciated Harm, Annals of Internal Medicine; 3 April 2012, vol. 156, no. 7

WEBSITES

www.uspreventive servicetaskforce.org

Survey Data on Acrylamide in Food: Total Diet Study Results for 2006:

http://www.fda.gov/Food/FoodSafety/FoodContaminantsAdulteration/ChemicalContaminants/Acrylamide/ucm053566.htm#table4

Photo courtesy of Wendy Newman Photography

CHARLES C. HARPE, MD

Dr. Harpe graduated from the University of Miami School of Medicine and completed his post-graduate training in internal medicine at the LDS Hospital in Salt Lake City, Utah before entering private practice as a general internist in 1996. Dr. Harpe is certified by the American Board of Internal Medicine. His service work in public health includes several years at the Davis County Health Board and later serving the Utah Partnership for Healthy Weight. Discovering health within the context of a person's cultural environment and socio-biological evolution is Dr. Harpe's most intellectually gratifying line of work these days.

www.ingramcontent.com/pod-product-compliance
Lightning Source LLC
Chambersburg PA
CBHW031148270326
41931CB00006B/186